Andrej Brodnik · Jan Vahrenhold (Eds.)

Informatics in Schools

Curricula, Competences, and Competitions

8th International Conference on Informatics in Schools:
Situation, Evolution, and Perspectives, ISSEP 2015
Ljubljana, Slovenia, September 28 – October 1, 2015
Proceedings

 Springer

Editors
Andrej Brodnik
Faculty of Computer and Information Science
University of Ljubljana
Ljubljana
Slovenia

Jan Vahrenhold
Institut für Informatik
WWU Münster
Münster
Germany

ISSN 0302-9743   ISSN 1611-3349   (electronic)
Lecture Notes in Computer Science
ISBN 978-3-319-25395-4   ISBN 978-3-319-25396-1   (eBook)
DOI 10.1007/978-3-319-25396-1

Library of Congress Control Number: 2015950925

LNCS Sublibrary: SL1 – Theoretical Computer Science and General Issues

Springer Cham Heidelberg New York Dordrecht London

Printed on acid-free paper

Springer International Publishing AG Switzerland is part of Springer Science+Business Media
(www.springer.com)

Lecture Notes in Computer Science 9378

Commenced Publication in 1973
Founding and Former Series Editors:
Gerhard Goos, Juris Hartmanis, and Jan van Leeuwen

More information about this series at http://www.springer.com/series/7407

Preface

This volume contains the papers presented at the 8th International Conference on Informatics in Schools: Situation, Evolution and Perspectives (ISSEP 2015). The conference was held at the University of Ljubljana, Slovenia, from September 28 to October 1, 2015.

ISSEP is a forum for researchers and practitioners in the area of informatics education, in both primary and secondary schools (K–12 education). It provides an opportunity for educators to reflect upon the goals and objectives of this subject, its curricula and various teaching/learning paradigms and topics, possible connections to everyday life, and various ways of establishing informatics education in schools. This conference also has an interest in teaching/learning materials, various forms of assessment, traditional and innovative educational research designs, the contribution of informatics to the preparation of individuals for the 21st century, motivating competitions, and projects and activities supporting informatics education in schools.

The ISSEP series started in 2005 in Klagenfurt, with subsequent meetings held in Vilnius (2006), Toruń (2008), Zürich (2010), Bratislava (2011), Oldenburg (2013), and Istanbul (2014). The 8th ISSEP conference was hosted by the University of Ljubljana, Faculty of Computer and Information Science.

The conference received 36 submissions. Each submission was reviewed by at up to four Program Committee members and evaluated on its quality, originality, and relevance to the conference. Overall, the Program Committee wrote 106 reviews. The committee selected 14 papers for inclusion in the LNCS proceedings, leading to an acceptance rate of 38.9%. The decision process was made electronically using the EasyChair conference management system.

In addition to the accepted contributions, this volume also contains abstracts of the invited lectures by Tim Bell (Christchurch), Maria Knobelsdorf (Hamburg), and Miha Kos (Ljubljana).

ISSEP was federated with a teacher conference for K–12 teachers. The conference was geared toward teachers from Austria, Italy, and Slovenia, although teachers from other countries also participated. The decision to federate the teacher conference and ISSEP was made so as to bring the results of computer science education research closer to the practising K–12 teachers. Moreover, since the participation at the teacher conference was international, the conference also provided a forum for the international exchange of ideas and experiences.

We would like to thank all the authors who responded to the call for papers, the invited speakers, the members of the Program Committee, the external reviewer, and – last but not least – the members of the Organizing Committee.

August 2015

Andrej Brodnik
Jan Vahrenhold

Organization

Program Committee

Erik Barendsen	Radboud University Nijmegen and Open Universiteit, The Netherlands
Andrej Brodnik (Chair)	University of Ljubljana and University of Primorska, Slovenia
Michael Caspersen	Aarhus University, Denmark
Valentina Dagiene	Vilnius University, Lithuania
Barbara Demo	Università di Torino, Italy
Ira Diethelm	Carl-von-Ossietzky-Universität Oldenburg, Germany
Kathi Fisler	Worcester Polytechnic Institute, UK
Yasemin Gülbahar	Ankara University, Turkey
Juraj Hromkovič	ETH Zürich, Switzerland
Peter Hubwieser	Technische Universität München, Germany
Peter Micheuz	Alpen-Adria-Universität Klagenfurt, Austria
Ralf Romeike	Friedrich-Alexander-Universität Erlangen-Nürnberg, Germany
Jože Rugelj	University of Ljubljana, Slovenia
Carsten Schulte	Freie Universität Berlin, Germany
Chris Stephenson	Google
Maciej M. Sysło	Nicolaus Copernicus University Toruń, Poland
Josh Tenenberg	University of Washington, USA
Françoise Tort	Ecole Normale Supérieure de Cachan, France
Jan Vahrenhold (Chair)	Westfälische Wilhelms-Universität Münster, Germany

Posters

Matija Lokar	University of Ljubljana, Slovenia

Workshops

Peter Micheuz	Alpen-Adria-Universität Klagenfurt, Austria

Teacher Conference

Mojca Bernik   University of Maribor, Slovenia
Barbara Demo   Università di Torino, Italy
Claudio Mirolo   Università di Udine, Italy
Peter Micheuz   Alpen-Adria-Universität Klagenfurt, Austria

External Reviewer

Filiz Kalelioğlu   Ankara University, Turkey

Organizing Committee

Andrej Brodnik   University of Ljubljana, Slovenia
Boštjan Borič   University of Ljubljana, Slovenia
Gašper Fele-Žorž   University of Ljubljana, Slovenia
Matevž Jekovec   University of Ljubljana, Slovenia
Nataša Mori   University of Ljubljana, Slovenia

Sponsoring Institution

University of Ljubljana, Faculty of Computer and Information Science

Invited Lectures
(Abstracts)

Surprising Computer Science

Tim Bell

University of Canterbury,
Christchurch, New Zealand
tim.bell@canterbury.ac.nz
http://www.cosc.canterbury.ac.nz/tim.bell/

Abstract. Much of what we can do with Computer Science seems like magic, such as searching billions of items in a fraction of a second, or decrypting a secure message without needing to know the key that was used to encrypt it. Other parts are surprising — surely given a fast enough computer we can find the optimal solution to a problem? This paper investigates magical and paradoxical ideas in computer science, and how these relate to Computer Science education.

The Theory Behind Theory - Computer Science Education Research Through the Lenses of Situated Learning

Maria Knobelsdorf

Universität Hamburg, Computer Science Department
Vogt-Kölln-Straße 30, 22527 Hamburg, Germany
knobelsdorf@informatik.uni-hamburg.de

Abstract. This paper introduces key characteristics of the situated learning approach and discusses from that perspective questions of pedagogy and educational research in Theory of Computation. This discussion exemplifies how a change in learning theories alters the unit of analysis, thus reframing research questions and potential answers. In its conclusion, this paper provides an outlook on potential research questions in secondary Computer Science Education.

Doubtology - About Common Sense, Doubt and Critical Thinking

Miha Kos

Ustanova Hiša eksperimentov, Trubarjeva ulica 39, 1000 Ljubljana, Slovenia

Abstract. Our society is facing a pandemic illness without a name but with clear symptoms: apathy in place of passionate curiosity, looking for a quick and easy way to learn instead of striving for in-depth knowledge, being compliant and conformant instead of thinking critically. This talk will focus on curiosity and critical thinking, two of the most important driving forces behind the learning process. What are we doing wrong during education which often seems to be stifling curiosity instead of nurturing it? How do we excite the power of imagination and curiosity and light the spark that will cause students to learn by themselves? Feeling curious already?

Throughout almost twenty years of experience running the Science Centre, I gradually realized that it is not a centre of science that I am running. Promotion of Science is just one of the tools used in order accomplish our main mission – inspiring curiosity and critical thinking.

Our society is facing a pandemic illness without a name but with clear symptoms of apathy in place of curiosity and learning, looking for easy but shallow ways of acquiring knowledge, no interest in seeking the answers to bothering questions, misinterpretation of dialog as being just two monologues, believing instead of having doubts, critical thinking, checking and proving...

There isn't a single country that would claim their educational system is perfect, or even that it is good. Experts are wondering when and where in the education process this curiosity is lost.

Curiosity is something every single human being is born with. Not only humans, many animals start their lives being curious. Curiosity is a driving force for the learning process. It is fuel for the trial and error process – learning by mistakes that are nothing more than personal learning experiences. Instead we have an educational system that despises mistakes rather than looks at them as a necessary learning optimisation method and encouraging them.

Imagine a curious child raising a hand in order to get the attention of the teacher and ask a question, pose a proposition or express a personal idea of the topic. This is one of the crucial moments that will define the future level of curiosity of the whole class.

There is a right and a wrong course of action. The teacher could respond with: "That's a good question/idea! Let's talk about it." "I don't know the answer. Does anyone have any ideas?" "Wow, great idea. What if we also take into account that ..." "This is a question that also bothered great scientists at that time." ...

The other response (that demands less effort) might be: "Don't interrupt the class!" "You should know that by now!" "What a crazy/stupid idea." "We will talk about it later." "We have already discussed this. Listen more carefully next time!" "Can someone please explain the idea to him/her. I am tired of repeating the same thing all over again and again." ...

One can guess which option is a "curiosity multiplier" and which is the "curiosity killer". Both options directly signal to the curios person (but also to the whole class) the value of being curious, but one option is treating this curiosity as a virtue (the holy grail for creativity) while the other shows that the curiosity does not pay, that the curiosity is punished.

Curiosity is a very tangible substance that each teacher should nurture throughout life. It triggers the passion for learning and creativity. It is also important as a teacher to exercise the answer "I don't know". It is precious to admit the mistakes one makes while teaching (especially if the teacher is alerted to the mistake by some doubtful student). This shows that everyone makes mistakes. Moreover it gives the teacher the feedback that 1) students are curious, 2) that they do care what the teacher is communicating, 3) that they don't just believe what they hear, 4) that they know to doubt and to think critically.

A good teacher is not a teacher at all. A good teacher is an inspirer that amplifies curiosity and encourages doubts and critical thinking. Inspired students will learn by themselves.

Contents

Surprising Computer Science

Tim Bell

University of Canterbury,
Christchurch, New Zealand
tim.bell@canterbury.ac.nz
http://www.cosc.canterbury.ac.nz/tim.bell/

Abstract. Much of what we can do with Computer Science seems like magic, such as searching billions of items in a fraction of a second, or decrypting a secure message without needing to know the key that was used to encrypt it. Other parts are surprising — surely given a fast enough computer we can find the optimal solution to a problem? This paper investigates magical and paradoxical ideas in computer science, and how these relate to Computer Science education.

Keywords: Computer science education, magic, paradoxes, fraud.

1 Introduction

It was only one generation ago that a computer was something that many people had no access to, and those who did have access perhaps shared one for a whole household, or queued up to use one in their place of work. Now it is common for each of our students to have multiple computing devices that are queuing up to be used — perhaps a desktop computer at home, a smartphone in their pocket, a laptop or tablet in their bag, a computer in a lab at school, and even a few old devices lying around that are no longer used or valued. With this transition has come the development of computing into a highly competitive consumer market where devices and software (typically "apps") are purchased through streamlined systems that enable consumers to keep up with the very latest offerings.

The arrival of very compact yet powerful computing devices such as smartphones and tablets, along with services such as search engines, social networks and online media, have created a digital ecology where the device and the software become impenetrable to the user; in fact, most devices would have their warranties voided if a student tried to look inside them, and online services protect themselves from letting users understand how they work, let alone being able to modify them. The opacity of such systems has essentially immunised a generation from believing they could understand what is going on behind the scenes. The curious might wonder how it is possible to search billions of web pages in a fraction of a second, or to accept hundreds of hours of video uploads every minute, or store thousands of songs in a device that weighs the same as a few coins. This puts the hardware and software of the digital revolution into

© Springer International Publishing Switzerland 2015
A. Brodnik and J. Vahrenhold (Eds.): ISSEP 2015, LNCS 9378, pp. 1–11, 2015.
DOI: 10.1007/978-3-319-25396-1_1

the realm of magic — we've seen it happen with our own eyes, but have no explanation for how it might work.

As teachers we can use the magic to get students excited, and our teaching can be thought of as showing students how the magic works. It is also important to help new teachers get beyond the magic and engage with the great ideas behind it.

In this paper we explore ideas from computer science that are indeed magic, but look at how we can reveal the magic to students, at the same time not wanting to lose the fascination of the discipline. We begin by reflecting on the ambiguously named area of "coding," and then look at how computer science has literally been used as the basis of magic tricks. From here we explore some paradoxes and surprises that arise in the discipline, and then look at how some people have used the mystery to perpetrate frauds. The conclusion discusses the lessons this provides for computer science education.

2   The Secret Code

If digital systems remain a mystery, then our main experience with them will be as *users* rather than *builders*. Lee *et al.* point out that students can engage with technology much better if they transition from *using the technology*, to *modifying it*, and *creating* new artefacts [13]. This reflects Rushkoff's "Program or be programmed" theme [14]: if we are not empowered to create new technology then we are doomed to fit in with whatever is created for us by technocrats.

From an economic point of view, any society that trains its children to be users is doomed to pay for products created by others. With regards to citizenship, if we are uninformed users then we are locked into accepting whatever technologies are made available to us, unless we are prepared to reject new developments altogether. Schulte and Knoblesdorf highlight the power of the belief that people have that they might never understand how computers work by drawing an analogy with the "muggles" (people without magical powers) in the Harry Potter stories [15]; the challenge is to convince people that they can develop these powers, and operate as an insider. Given the growing awareness that students need to engage with technology, it's not surprising there has been a movement back to bare-bones systems such as the Raspberry Pi [17], which invite the user to modify and create, rather than just use.

As more countries are introducing computer science and computational thinking into their curriculum, teachers are having to overcome fear and confusion that they have around the digital world so that they can teach areas such as computer science and computational thinking to their students in a way that is authentic and engaging. Much of the value of having ways to reveal the "secrets" behind the technology is that it is important for empowering new teachers to feel that they could teach this important discipline.

The recent growth of interest in the topic of "coding" (used as a buzz word for programming) has had a role in opening up this mystery to the public. The word "code" is mysterious and has several meanings in computer science, which all adds to the confusion for outsiders. It conjures up the idea of secret codes, and the public may have come across the term in contexts such as the Enigma codes or the fictional Da Vinci code. There is a sense that for such codes, once the secret key is known then everything is revealed. Organisations that have worked to help students learn to code give them the opportunity to break into this mysterious world, and having the opportunity to engage with this prior to a student's teenage years is valuable [6]. Of course, learning to "code" doesn't really unlock *all* the secrets, but it can be a powerful enabler that helps young students to engage with the basic principles and get over the initial hurdle of not knowing what programming is. In principle, learning just the basics of any Turing-complete language (e.g. selection, iteration and variables) is sufficient to be able to write any program. but understanding programming properly takes years,[1] and given the development of new languages, there will always be more to learn.

This sense of the word "coding" is rather specific. Even in the limited context of software development, traditionally the part of programming that is "coding" might be the most routine part, simply converting a design to code. In contemporary contexts it often refers to the whole process of designing, implementing, testing and debugging software. Thus even when referring to programming, "coding" has acquired two meanings, much like the word "hacker" can be a positive reference to broad technical prowess, but can also have a strongly pejorative sense.

To add to the confusion, the term "code" is ambiguous and pervasive in computer science. When applied to data, the term "code" appears in source coding, channel coding, and cryptographic coding. Each of these are about representing data to transmit and store it efficiently, reliably, and securely. To add to the confusion, in computing we have "source code," which is quite different to "source coding!" And then source code is compiled to machine code, which is also referred to as the object code, or even as the "binary" of a program.

This leads us to the most widespread code in computing: "binary". It is widespread because it is the fundamental representation in digital devices — the word "digital" is often used to describe the new technology (e.g. "digital revolution"), and the term primarily refers to the binary digits that are fundamental to all computing devices. This brings its own paradox — if you ask a group of students if they are aware that computers store only zeroes and ones, this is likely to elicit a positive response, yet the truth is there are no actual zeroes or ones stored in a computer. There are high and low voltages in memory, pits and lands on optical disks, high and low pitches on modems,[2] black and white stripes on bar codes, but no zeroes or ones! The bit is an abstract representation of a physical phenomenon; physical bits use space and energy.

---

[1] Norvig discusses this in his essay "Teach yourself programming in ten years" (http://norvig.com/21-days.html).

[2] To be accurate, modems often use four or more sounds, and other methods to encode multiple bits.

Ultimately every code mentioned in this section is eventually represented in binary, and hence understanding binary code can be a powerful enabler for students of computer science, and indeed for the general public. It is a common topic in computing curricula, and there are many resources for teaching it (a survey of games for teaching CS found more resources for teaching binary than any other topic [8]), but often it is taught in a very limiting way, focussing on how to convert between decimal and binary. This could give the impression that computer scientists spend a lot of time doing these conversions, but the real point is to understand the power — and limitations — of this representation. For example, a 16-bit number doesn't have twice the range of an 8-bit number; 24-bit colour isn't three times better than 8-bit colour; and a 2048-bit encryption key isn't just twice as strong as a 1024-bit key (in fact, in principle a 1025-bit key is twice as good as a 1024-bit key).

Binary representations have relevance when matched to human needs: humans can perceive millions of colours, so an 8-bit or 16-bit palette isn't sufficient to out-perform human perception; we can detect time delays of 0.1 seconds, so a 10Mbps connection will introduce a perceivable delay if downloading a one megabyte photo; and someone attacking an encryption key may find the data useful if cracked in less than a day, so a 56-bit key is not sufficient.[3]

The basics of binary representation can be taught very simply, and to large groups, using the CS Unplugged activity with around 5 to 8 cards (one per bit), with each card having the corresponding value visible, or not visible. Figure 1 shows eight students holding such cards to show 179 dots. Asking the audience if each card is needed to display a number elicits yes/no responses that illustrate how a value can be communicated using just two distinct signals. Asking questions such as "what is the largest number they can represent," "what is the smallest number they can represent," and "how often is the right-hand bit flipped when counting" enables the audience to engage with the patterns around binary numbers. For the students holding the cards, it is an opportunity to find out how simple a bit is!

A key point about digital representations is that they are fundamentally different from analogue representations. Because of channel coding (error correction) information can be stored and transmitted with an expectation of an exact copy being retrieved; because of source coding (compression) this can be done in sufficiently little time and space that humans find it useful (e.g. storing thousands of songs on a pocket size device); and because of encryption coding the data can be stored and transmitted on public systems with good assurance of privacy for the data.

Binary may appear to be a secret code to outsiders, but it is easily explained, and unlocks the ability to conceive what is actually happening in our digital world.

---

[3] Until 1999, the US government regulated encryption keys stronger than 56 bits. In 1999, a 56-bit encrypted code was cracked in less than one day (https://en.wikipedia.org/wiki/DES_Challenges).

Fig. 1. Students acting as bits to illustrate binary representation to an audience

3 Communicating the Magic

As well as the mystery of "code", much of what we can do with Computer Science seems like magic, such as searching billions of items in a fraction of a second, or sharing a secure message without having to send the key that was used to encrypt it. It is valuable to be able to share this magic with an audience in a relatively short time to help them see that it can be understood. The discipline (magic) behind computing is commonly mixed up with how to *use* computers (e.g. digital literacy, or the vague term "ICT" [7]), so we need to be able to clearly show the difference, whether talking to key influencers (including parents, grandparents, or education officials) or the students themselves.

Many presenters have done this by literally using magic tricks based on ideas from computing. This is a useful approach because it impresses the audience and creates a lot of motivation to find out how it works. When the secret is revealed, it empowers the audience as they understand the concept.

Examples of magic being used to expose computer science to the public include a show based on CS Unplugged material [2], the Villanova magic schools [9], the CS4FN magic shows [5] including books on "The Magic of Computer Science[4]", public lectures on the "wonders of informatics" [10], books based on analogies with magic and fairy tales (e.g. Kubica's "Computational Fairy Tales" [11] and "Best Practices of Spell Design" about programming [12], and Bueno's "Lauren Ipsum" [4]), and the Aachen Infosphere "Zauberschule Informatik[5]" (magic school of computer science). The idea that programming is creating something out of thoughts has also been used to describe it as magic (e.g. by Werner *et al.* [18] and in the CSTA K-12 standards [16]), and there are even programming lessons based on creating "spells" (e.g. the Code Spells challenge, http://codespells.org/).

---

[4] Available for free download from http://www.cs4fn.org/magic/magicdownload.php
[5] http://schuelerlabor.informatik.rwth-aachen.de/modul/zauberschule-infor
matik-ein-erster-einblick-die-welt-der-informatik

Each of these approaches engages an audience by creating mystery, and then revealing how a clever idea or concept can be used to give the illusion of some kind of magical power. They are provided in formats that make them accessible to an audience that might be sceptical, and isn't prepared to invest a lot of time or effort into understanding computer science, possibly based on their incorrect belief that it would be a waste of time for them to try.

4   Paradoxes and Surprises

Some parts of computer science are simply paradoxical or surprising. Creating an awareness of this can help students to take a more curious view of phenomena surrounding computing, and help to demonstrate how there's a lot more to computer science than learning to program.

For example, a student might believe that given a fast enough computer we can find the optimal solution to a problem. An example that shatters this idea is the travelling salesman problem, where (for example) a courier must visit a number of houses to drop off parcels, and they would like to minimise the time and fuel used to do this.

Figure 2a shows the problem being solved by brute force, evaluating every possible route, for seven locations. This version takes about half a minute to find the solution. After showing it to students, we can ask how long they think it would take to solve the problem for twice as many locations. Typically they will estimate that it is twice as long (about a minute), and are surprised to find out that it is actually several years (Figure 2b). The surprise is easily explained by considering the combinatoric explosion of possibilities. However, the impossibility of the approach becomes clearer as the number of locations is increased further, to the point where even the fastest possible computer would take centuries to solve it for a relatively small number of locations.

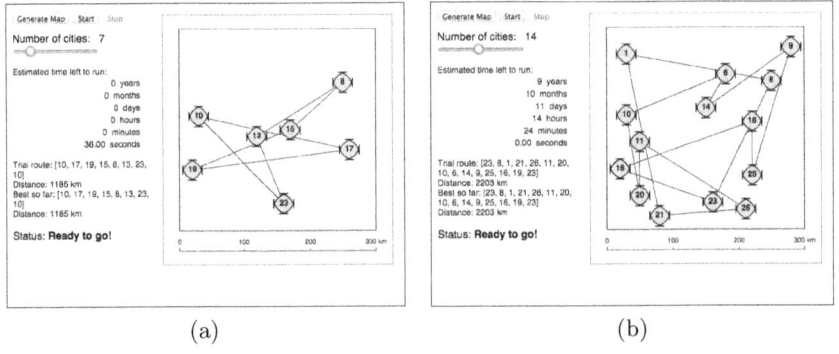

(a)   (b)

Fig. 2. Solving the TSP for (a) 7 locations and (b) 14 locations

The surprise occurs in the opposite direction when considering binary search. Figure 3 shows a student searching for a number hidden under one of 15 cups;

to look under each cup they must surrender a lollie (candy), and they have only 5 lollies to start with. This seems futile, but by using a binary search (which students will often work out for themselves), they can eliminate half of the cups with each probe. Given that 4 lollies is sufficient to search 15 cups, the students are then asked how many would be needed for 30 cups. The natural reaction is that it would be twice as many, but they soon realise that only one more probe is needed to cope with twice the number of items. Extending this, students can work out how easy it is to search 1000 cups (10 probes), a million cups (20 probes), or even 1,000,000,000 cups (30 probes). At this point it becomes clear that a search engine that can search billions of items in a fraction of a second isn't so magic — you just need to use the right algorithm.

Fig. 3. Searching for a value hidden under cups

These two examples show a simple concept: that the time taken by a program isn't necessarily proportional to the amount of input. These ideas help us return to the magic and beauty of computer science — they aren't tricks, but are just the way things are if you work through the concepts. Things might not be as expected, but we can understand and predict them if we think it through.

Other paradoxes and surprises that come up include:

- one of the fastest sorting algorithms (quicksort) is slowest when given a list of identical items (unless you take special measures);
- for NP-complete problems, a child could design a small problem that they know the solution for (e.g. the dominating sets problem [3], presented in an unplugged format as the "Tourist Town" problem[6]), but a computer would take billions of years to solve it using even the best known algorithms;
- randomness can help make algorithms run faster (for example, hashing is best when the function is random);

---

[6] http://csunplugged.org/dominating-sets/

- finding the shortest path in a graph is easy (and regularly used by GPS devices), yet finding the longest path is NP-hard and therefore no good solution is known;
- lossless compression expands more files than it reduces;
- secure password checking systems don't store passwords, yet they can check if you've entered the right password; and
- you can set up a secure communication encryption code in front of an eavesdropper, who knows every detail of the method and all the information you have exchanged with your friend, and yet they can't make sense of the communication between you and your friend (Public-key cryptography).

All of these are concepts that can be exposed to relatively young students with the right scaffolding, and yet help us to maintain the intrigue of the subject.

5   The Dark Side of Magic

Because of the digital revolution we have been through, society has become used to new inventions that seem too good to be true: web sites that store billions of videos, devices that can locate where an overseas friend is within a few metres, web sites that automatically label photos with the names of the people in them, and systems that can send spare parts through the internet, to be printed where they are needed. The phrase "too good to be true" is also used as a warning that something might be a fraud, and by seeing technology as magic, users have become desensitised to this possibility. A common example is that millions of spam emails can be generated with a few lines of program code, but to the user it appears that they're receiving an email specific to them, for example, from their own bank or ISP, unaware that thousands of others have received the same email, but for them it was the wrong bank. This illusion is now well known, and yet sufficient people fall for it that there are still billions of spam emails sent every day. Understanding how technology works helps users to avoid being caught out.

Another example of fraud that has succeeded from ignorance of computing principles is a series of startup companies that each claimed to be able to compress any file by a significant amount, and later reconstruct the original exactly. Examples include the "Wider Electronic Bandwidth" company announced in Byte magazine in June 1992 [1], "Adam's Platform" (1998), Madison Priest's "Magic Box" (c1994), and the "NearZero" system (2001).

In the case of "NearZero", the claim was that *any* file could be compressed to about 7% of its original size. As with the others, demonstrations were given to potential investors showing files being replaced with very small compressed versions, which were then expanded back to their original size. Since this would speed up networks and increase disk storage by a factor of 15, investors flocked to get in early on the system, and many millions of dollars were sunk into the company. None of the systems ever resulted in a commercial product, and a lot of people were left out of pocket.

There are several explanations for why such a system isn't possible; one is that if *any* file can be compressed then its own compressed files can be compressed, and thus any file could be reduced to one byte (or even one bit!), which doesn't allow for many files to be represented. The claim could be exposed by providing 257 different 16-bit files, and asking to have them all compressed to 1 byte; two of them will have identical representations and can't be decompressed accurately. A similar argument (based on the pigeon-hole principle) can be used to show that every lossless compression method must expand at least as many files as it reduces.

The demonstrations themselves involved the digital equivalent of a magician's smoke and mirrors: the compression program would typically copy the file being compressed to an unused part of the disk and replace it with the "compressed" file. The decompression process involved simply copying it back, with suitable delays introduced to make it look like some work was being done. The viewer was misdirected by having them focus on the compressed file, making them oblivious to the disk usage going up because of the copied file appearing in a hidden location.

This is a somewhat extreme example of the general public being duped out of millions of dollars because of a lack of understanding of basic principles, in this case, data representation. However, there are many important decisions that people make around digital technology; issues like our privacy, our ability to verify authenticity of online interactions, and the ability of organisations and governments to monitor individuals are all heavily impacted by issues such as the encryption, storage, and transmission of data. It is important for society to have some understanding about the technologies that permeate our lives and relationships, in the same way that understanding the science behind other innovations (such as genetic modification or nuclear power) enables us to have informed opinions on their benefits and risks.

6 Conclusion

The view of digital systems as magic can be both disempowering (if it can't be understood, it's not worth trying) and exciting (it's a mystery worth solving). Our challenge is to help students — and more urgently, teachers — overcome the view that it is so magic that they couldn't understand it, yet still retain the fascination that makes it an attractive discipline that is full of surprises and mysteries. As countries adopt computer science as a mainstream topic in schools, the largest hurdle they are facing is preparing teachers for this new discipline. This means that enabling teachers to see both the magic of the topic, and that they can also understand the magic, is crucial to the success of new curricula built around how to be a creator of technology, rather than a user. Getting students, and teachers, past the initial hurdle so that they have some insight into how the magic works is a vital step, and learning computational thinking and programming early with the right tools empowers students to take on bigger challenges.

Perhaps there is a concern that revealing the tricks behind the magic will remove the fascination of the subject. Sometimes in a magic show the audience is told how a trick works so they can see that they could do it themselves. Then, just when they think they can understand it, the magician takes the trick one step further and does something that once again seems impossible. The beauty of computer science is that there is always one more trick, one more challenge, one more unsolved problem. The world of digital technology has infinite possibilities, and with creative people developing new ideas, there is no risk that we will lose the sense of magic around this exciting field. The challenge is to get them started.

Acknowledgments. The author is grateful to James Atlas and Caitlin Duncan for valuable input to this paper. The photographs are courtesy of Sam Jarman and students at Chisnallwood Intermediate School, taken by Jack Morgan and Gerard MacManus.

References

1. Microbytes: instant gigabytes? Byte Magazine 17(6), 45 (1992)
2. Bell, T.: A low-cost high-impact computer science show for family audiences. In: Australasian Computer Science Conference 2000 (ACSC 2000), Canberra, Australia, January 31-February 3, pp. 10–16 (2000)
3. Bell, T.C., Witten, H.I., Fellows, M.: Computer Science Unplugged: Off-line activities and games for all ages (original book) (1999), http://csunplugged.org
4. Bueno, C.: Lauren Ipsum: A Story About Computer Science and Other Improbable Things. No Starch Press (2014)
5. Curzon, P., McOwan, P.W.: Engaging with computer science through magic shows. In: Proceedings of the 13th Annual Conference on Innovation and Technology in Computer Science Education, ITiCSE 2008, pp. 179–183. ACM, New York (2008)
6. Duncan, C., Bell, T., Tanimoto, S.: Should your 8-year-old learn coding? In: Proceedings of the 9th Workshop in Primary and Secondary Computing Education - WiPSCE 2014, pp. 60–69. ACM Press, New York (2014)
7. Furber, S. (ed.): Shut down or restart? The way forward for computing in UK schools. The Royal soceity, London (2012)
8. Gibson, B., Bell, T.: Evaluation of games for teaching computer science. In: The 8th Workshop in Primary and Secondary Computing Education (WiPSCE 2013) (2013)
9. Hess, K.L., Papalaskari, M,-A., Weinstein, R., Styer, R., Way, T., Lagalante, A.: Special session-creation of the Milwaukee School of Magic. In: 37th Annual Frontiers in Education Conference-Global Engineering: Knowledge Without Borders, Opportunities Without Passports, FIE 2007, pp. S1F–1. IEEE (2007)
10. Hromkovic, J.: Algorithmic Adventures: From Knowledge To Magic. Springer, Heidelberg (2009)
11. Kubica, J.: Computational Fairy Tales. CreateSpace Independent Publishing Platform (2012)
12. Kubica, J.: Best Practices of Spell Design. CreateSpace Independent Publishing Platform (2013)
13. Lee, I., Martin, F., Apone, K.: Integrating computational thinking across the K-8 curriculum. ACM Inroads 5(4), 64–71 (2014)

14. Rushkoff, D.: Program or be programmed: Ten commands for a digital age. OR Books (2010)
15. Schulte, C., Knobelsdorf, M.: Attitudes towards computer science – computing experiences as a starting point and barrier to computer science. In: Proceedings of the Third International Workshop on Computing Education Research – ICER 2007, p. 27. ACM Press, New York (2007)
16. Seehorn, D., Carey, S., Fuschetto, B., Lee, I., Moix, D., O'Grady-Cunniff, D., Owens, B.B., Stephenson, C., Verno, A.: CSTA K-12 Computer Science Standards: Revised 2011. Technical report, New York (2011)
17. Upton, E., Veloso, M., Gadgil, A., White, T., Monks, B., Malkin, R.: Today's engineering heroes. IEEE Spectrum 52(3), 39–49 (2015)
18. Werner, L., Denner, J., Campe, S., Kawamoto, D.C.: The fairy performance assessment. In: Proceedings of the 43rd ACM Technical Symposium on Computer Science Education - SIGCSE 2012, p. 215. ACM Press, New York (2012)

The Theory Behind Theory - Computer Science Education Research Through the Lenses of Situated Learning

Maria Knobelsdorf

Universität Hamburg, Computer Science Department
Vogt-Kölln-Straße 30, 22527 Hamburg, Germany
knobelsdorf@informatik.uni-hamburg.de

Abstract. This paper introduces key characteristics of the situated learning approach and discusses from that perspective questions of pedagogy and educational research in Theory of Computation. This discussion exemplifies how a change in learning theories alters the unit of analysis, thus reframing research questions and potential answers. In its conclusion, this paper provides an outlook on potential research questions in secondary Computer Science Education.

1   Introduction

In the research community of Computer Science Education (CS Ed), the awareness for discussing and explicitly incorporating theoretical frameworks into research is constantly rising [17], [6]. Theories and concepts of how individuals learn play an important role in educational research because they not only affect which research questions are posed and what kind of data collection and analysis methods are chosen, but more importantly influence the development of pedagogical concepts and interventions. While cognitivist and constructivist concepts of learning are established frameworks in CS Ed, recent theories and related discourses from educational psychological research are still being introduced and discussed. Theories that go under the names *situated learning* [17], *sociocultural learning* [23], *situated cognition theory* [3], *distributed intelligence* [21], or *activity theory* [11] have been developed over the last three decades and started to become more important in field of CS Ed research, see for example [2], [10], [14], [16], [22], [26]. Likewise, in other educational disciplines the situated cognition theory became influential [19], which also inspired the development of comparable approaches for secondary CS Ed [7], [13]. These new approaches extend and challenge established beliefs and understanding of learning and therefore the corresponding research programs and their achievements in related pedagogies and didactics.

In this paper, I summarize key characteristics of the situated learning approach and discuss from that perspective questions of pedagogy and educational research in CS Ed. For this matter, I rely on previous work, i.e., especially [26], a review of sociocultural cognition theory, as well as [15], [13], [14], where parts of concepts and arguments presented in this paper have been already introduced and discussed.

© Springer International Publishing Switzerland 2015
A. Brodnik and J. Vahrenhold (Eds.): ISSEP 2015, LNCS 9378, pp. 12–21, 2015.
DOI: 10.1007/978-3-319-25396-1_2

In particular, I summarize modifications to the pedagogy of an undergraduate Theory course held at the University of Potsdam, Germany. Here, I contributed to by taking into account the pedagogical approach of cognitive apprenticeship which led to a significant reduction of the course's failure rates in the final exam [15]. I will reflect these changes from the perspective of situated learning and draw conclusions for further research investigations in the educational scope of Theory of Computation.

The situated learning perspective represents a paradigm change from many other kinds of psychological frameworks [25]. Some of the concepts do not have straightforward mappings to established psychological theories, and must be understood as part of a larger, but different, theoretical whole. Such a paradigm shift in psychological theory may engender the kinds of cognitive dissonance for the readers that are also felt by an experienced imperative programmer on first encountering an object-oriented language. Therefore, readers with a strong background in CS and cognitive theories may find it challenging to adopt and appreciate this way of thinking. However, the useful new lenses that the approach offers is worth the endeavor because it significantly broadens our understanding of learning on which sustained innovation research for CS Ed can unfold.

2   Theoretical Framework

2.1   The Situated Learning Perspective

The cognitive view on learning has its roots in cognitive psychology and artificial intelligence research. Learning is conceptualized as a process in which individuals create a mental representation of a specific knowledge entity in their minds. A person's cognitive processes operate on such mental models, are based on logic-like rules of inference, and are understood to happen solely in the person's mind. Hence, the approach focuses on the question how specific knowledge is acquired and represented in the mind of an individual ([26], p. 5-6). The cognitive approach was criticized for being too much focused only on the single student and solely on his or her cognitive process while neglecting the social and cultural environment in which students' learning takes place [1], [3], [7].

The situated learning approach has its roots in Russian cultural-historical psychology developed by Vygotsky [27] and was strongly influenced by insights from artificial intelligence, as well as developmental psychological research ([26], p. 5-6). From a situated perspective, learning is conceptualized as a process of enculturation into a domain-specific community. The latter is often called "Community of Practice" (CoP), a term coined by Lave and Wenger [17], in order to emphasize "practices in which individuals have learned to participate, rather than on knowledge that they have acquired" ([7], p. 8). From this perspective, the goal of learning is to become a full member of a CoP.

In the trajectory of participation, individuals of a CoP learn to interact with each other mediated by material and representational systems, which are often metaphorically referred to as *tools*. In this context, the term "tool" denotes material objects such as pencils, hammers, automobiles, or coffee machines as well as representational systems such as "language; various systems for counting; mnemonic techniques;

algebraic symbol systems; works of art; writing; schemes, diagrams, maps, and mechanical drawings" and similar ([28], p. 137). Both aspects of a tool affect the material and mental activity of an individual in a CoP.

Tools do not simply arise de novo in the hands and minds of individual actors. Rather, they are provided to individuals by the surrounding culture of a CoP, accreting over time and, passed from one generation to the next. As Pea points out, tools "represent some individual's or some community's decision that the means thus offered should be reified, made stable, as a quasi-experiment form, for use by others. In terms of cultural history, these tools and the practices of the user community that accompany them are major carriers of patterns of previous reasoning" ([21], p. 53). Cultural practices of tool use evolve in tandem with the evolution of the tool. Therefore, tools represent socially distributed cultural entities that implicitly embed collective knowledge of their use ([26], p. 8-10).

2.2   Implications for Research and Formal Education

One of the most important implications for educational research is that the understanding and specific skills that students develop during an activity depend strongly on the tools used to carry out the activity. In consequence, mental processes, tool use, and interaction with the world are regarded to be tightly bound together: "This has the important implication that when understanding learning, we have to consider that the unit that we are studying is people in action using tools of some kind (see Wertsch, 1991, 1998; Säljö 1996). The learning is not only inside the person, but in his or her ability to use a particular set of tools in productive ways and for particular purposes." ([24], p. 147). In addition, Greeno points out that "[w]hen an analysis of an individual's knowing is proposed, the analysis should be an account of the ways that the person interacts with other systems in the situation. Just presenting hypotheses about the knowledge someone has acquired, considered as structures in the person's mind, is unacceptably incomplete, because it does not specify how the other systems in the environment (including other people) contribute to the interaction." ([7], p. 8).

Regarding formal education, a major critique is that parts of practices and tool use of a CoP are singled out, formalized, and organized around a curriculum with courses, assignments, and tests. This is depriving a community's practice of the coherent whole it represents within the community making it difficult for students to adopt the practice and become engaged. Lave and Wenger argue that learning in conventional schools "is predicated on claims that knowledge can be decontextualized" (p. 40) and "suggest that learning occurs through centripetal participation in the learning curriculum of the ambient community. Because the place of knowledge is within a community of practice, questions of learning must be addressed within the development cycles of that community" ([17], p. 100). Therefore, educational settings in school need to embed curriculum topics in authentic contexts in which students can better recognize the full meaning and reasons of specific practices and tools of a CoP.

Ben-Ari among others has pointed out that formal education of a specific domain has slightly different goals as well as constraints than traditional apprenticeship-like settings of education. Besides economical reasons, secondary but also tertiary education is designed to enable students broad participation in different domains or subfields of a domain and preclude a premature determination of future occupation ([2], p. 88). Students are supposed to enculturate into different domains enough to get acquainted with a broad range of practices from different disciplines or subfields and later on choose a specialization in order to fully grow into the CoP of their choice.

On the other hand, a school also represents a CoP with a specific culture, practices, tools, members, etc. and Greeno points out that "[m]ethods of instruction are not only instruments for acquiring skills; they also are practices in which students learn to participate. In these practices, students develop patterns of participation that contribute to their identities as learners, which include the ways in which they take initiative and responsibility for their learning and function actively in the formulation of goals and criteria for their success." ([7], p. 9). In consequence, when discussing educational settings and related research more attention needs to be paid how student enculturation takes place and how the curriculum, pedagogical approaches, teacher education, etc. are supporting students in this process.

Collins et al. argue that in traditional apprenticeship tasks or activities required to be accomplished by the students make sense for them because it is part of a coherent whole and arises from the demands of a specific workplace [5]. For this matter, educational settings need to put forward all aspects of practices and tools of a community and teach them with regard to students' enculturation process. Especially in higher education, where students are supposed to adopt the expertise of a particular scientific community, it is not enough paid attention "to the reasoning and strategies that experts employ when they acquire knowledge or put it to work to solve complex or real-life tasks. [...] To make real differences in students' skill, we need both to understand the nature of expert practice and to devise methods that are appropriate to learning that practice" ([5], p. 38-39).

In the next section, the situated learning approach and these implications will be discussed within the context of Theory of Computation.

3   Enculturation into Theory of Computation Community

At German universities, Theory of Computation is considered one of the fundaments of academic CS education and introductory courses to Theory of Computation are an integral part of undergraduate CS Ed programs. This is also the case at the department of CS at the University of Potsdam, Germany. The Theory of Computation courses cover the foundations of automata, programming languages, computability, and computational complexity. The corresponding concepts, theories, and algorithms are strongly mathematical in nature, regarding formalized inscriptions used for the discourse as well as a strong focus on mathematical proofing of presented theories and approaches (for better overview see for example [9]). By introducing idealized mathematical models of the computer and discussing methods for designing and analyzing them, students are supposed to develop the ability of thinking abstractly about computational processes and models. However, before the introduced modifications in the course's pedagogy many students had problems in achieving these goals and failure rates in final exams were very high (usually between 30-60%).

From the situative perspective, Theory of Computation is a specific CoP within the bigger community of CS. The Theory community includes members which mostly work at academic institutions like universities and therefore the discourse of the community is strongly based on academic research practices. In addition, the community's culture is strongly affected by mathematical practices and inscriptions. Academic education in this field can be regarded as the first step towards enculturation into this specific CoP. Since members of this community teach Theory courses at the university, consequently the pedagogy of these courses implicitly represents the community's beliefs and practices of how to enculturate newcomers. This includes the topics covered in these courses as well as goals and assumptions of how students are supposed to learn and work successfully. This needs to be taken into account when arguing for changes or improvements of established pedagogies in this field.

3.1   Teaching Practices: Theory of Computation

Until the course setup and its pedagogical approach were modified, the Theory course consisted of the following components, which are typical for an introductory CS course in Germany:

- Approximately 90-120 minutes of lectures per week given by a faculty member who presents the course topics, central concepts, algorithms, and their proofs and illustrates them with examples.
- Weekly homework assignments based on the current lecture topics, which students are expected to solve individually and submit in writing for reviewing and grading by tutors (usually senior students). Handing in homework can but doesn't need to be mandatory.
- Approximately 90 minutes student session per week attended by approximately 25-30 students and chaired by teaching assistants, during which students are expected to present their solutions to last weeks' homework assignment. The objective of these sessions is to give them an opportunity to check the correctness of their solutions and discuss them with the group.
- Summative assessment by the end of the course including several assignments to be solved in written form.

Within this pedagogical approach it was implicitly assumed that during the lectures students can follow and understand the presented concepts, theorems, and corresponding proofs, and are also able to deduce from the presented practices how to handle and work on the weekly homework assignments by themselves. In addition, student sessions were supposed to serve students as verification and improvement of their self-directed development of practices in Theory of Computation. However, teaching assistants reported that students had difficulties coping with the contents of the course due to its abstract and theoretical nature and that students seemed not to know how to tackle weekly homework assignments.

The situated learning theory drew attention to the Theory practices and how they were taught in the course. It was observed that these were not addressed and exposed sufficiently. Since all lecture topics were prepared in advance for a smooth presentation, students did not experience the enormous effort it took to create them, especially since the historical dimension of this field was not part of the curriculum. In consequence, students did not experience on a regular basis how the teacher or professor (who represents the expert member of the Theory of Computation CoP) is approaching a new problem, trying out different approaches, making mistakes, taking notes, etc. before creating a solution.

Altering the pedagogy of the course, the main idea was to use student sessions to demonstrate and discuss relevant techniques of how to solve specific problems so that students are better prepared to solve their homework by themselves (for more details see *scaffolding & fading* method [5]). Also, the textbook used in the course was chosen with regard to a strong emphasis on explaining and reflecting presented approaches [9].

Furthermore, important tools in Theory of Computation are mathematical inscriptions and visualizations both created with pen and paper or chalk and blackboard. These tools are used to describe, specify, and reason about ideas, approaches, examples and potential solutions. We found it important to demonstrate and explain students how these tools are meant to be used in this field of discourse. For this reason, we offered specific preparatory exercises that were solved jointly during the student sessions and served as a preparation for the homework assignments. These preparatory exercises included detailed written solutions with extensive reflections and explanations of the solution, providing insights about the used technique, method or strategy. Also, the solutions explicated how mathematical inscriptions are used in this field and are expected to be used by the students when they turn in their homework.

Redesigning the student sessions, the major change was to align the preparatory exercises with homework assignment in respect to structure and content. This means that in each session the same type and amount of problems was used for the exercises as well as for the homework assignments. In addition, it corresponds to students' activities during final exam, where students have to formulate a written solution for an assignment, which then will be graded by tutors under supervision of the instructors. In order to understand the expectations, especially with respect to the very strict and formal character of solutions to assignments, students need to train this skill and to receive a weekly feedback about their efforts. Since the final exam represents a situated activity particularly relevant of the overall pedagogy of the course, we believed that students must be prepared for this as well, especially when this is their very first university exam. For this reason, we also started to offer a *pre-exam* after the first half of the course. The pre-exam gave students an opportunity to practice the assessment situation and explicates what will be expected from them during the finals.

Students acknowledged the described pedagogical changes. In a survey conducted during the course, they reported to feel comfortable with the weekly homework and overall expectations of the course since they would know how to work on their assignments. As a result, the failure rate decreased 60% to below 10% while keeping the requirements for final exams comparable to those of the previous years. However, other CS colleagues, with whom we talked about this approach, were concerned that students are not fully acquainted with what they called "real" Theory of Computation. They argued that students are just trained to understand and apply presented solutions and are not learning to develop solutions to given computational problems on their own. This argument is very important because it emphasizes 1) how relevant the ability of *solving* computational problems is in the community of Theory of Computation in comparison to understanding given solutions and 2) the expectation that students should develop this ability without any scaffolding and from the very beginning of attending a Theory course. We argued that in order to be able to develop solutions to theoretical computational problems students should first understand and apply given solutions and the success of our students proved this approach to be right. However, there is no substantial empirical evidence showing how students best develop this important ability in Theory of Computation and what kind of pedagogy is required for supporting this adequately. Investigating this research question would require revealing the community's implicit beliefs about educational goals and pedagogical practices in this subfield of CS.

3.2   CS Students' Engagement in Theory of Computation

Since the majority of CS graduates do not pursue a research career in Theory of Computation, it can be concluded that except for a temporary enculturation, CS students do not intend to become full members of the Theory of Computation CoP. In addition, in the survey conducted during the Theory course at the University of Potsdam, most CS students reported that they did not find the course topics to be particularly interesting and only attended the   course because it was mandatory. When asked for detailed reasons, most students explained that the purpose of studying Theory of Computation was unclear to them. Therefore, it seems to beimportant to explicitly provide students with reasonable rationales for studying this field of CS in order to foster their interest in becoming more engaged in Theory practices.

One line of reasoning for including Theory of Computation into the CS curriculum, which is voiced regularly in the community,   is that prospective computer scientists should be familiar with the theoretical underpinning of computing. Students who intend to become members of the software engineering community need to be familiar with topics like complexity or design and analysis of algorithms. For this matter, short examples of a topic's possible applicability are usually provided during the lecture and this was also the case in the course at the University of Potsdam. The challenge is that CS students really start to understand this argument much later in their education and therefore require additional reasons to spark their interest when being newcomers to the Theory of Computation CoP.

In line with the purpose of enculturating students, presenting students with computational problems and teaching them practices of how to develop solutions and proof their correctness both emphasize the CoP's focus on these activates. In addition, it can be helpful to explain how theoretical computer scientists are motivated to work on computational problems. Addressing this goal, Chesñevar et al. [4] introduced "biographical notes, videos and articles associated with the historical context in which the theory of computing emerged as a new discipline" (p. 8). The authors reported that students responded very well to this historical perspective of theoretical CS, developed a deeper understanding, and became therefore more engaged. It seems that the historical perspective is a good approach not only to help students in creating meaning but also the teachers of the course to reflect and emphasize on the reasoning and strategies of creating this body of knowledge. Nevertheless, it is not yet clear if understanding the CoP's motivation will actually make it meaningful for students as well. In this context, another research question is how practices of Theory can contribute to students' practices of other CS fields, e.g., modeling or programming and – if there is evidence for such transfer – how this can be used to engage students in addition.

4   Conclusion and Outlook

This paper introduced key characteristics of the situated learning theory and discussed questions of pedagogy and educational research in the field of Theory of Computation from this perspective. This discussion exemplifies how the change in learning theories alters the unit of analysis, thus reframing research questions and potential answers. Situated learning theories focus on a unit formed by individuals, interacting with each other and with tools emphasizing patterns of participation and the culture of learning in this unit of analysis.

Discussing questions of learning and formal education in the field of Theory of Computation is exemplary because tertiary education in this field is usually implemented by the members of the related (scientific) Community of Practice (CoP). The trajectory of enculturation is intended to start with introductory courses to Theory of Computation and leads to advanced seminars with open problems from the field. On the other hand, introductory courses to programming are more complex regarding the questions of community belonging and enculturation. The latter is meant towards a CS community, but this is a shared roof of different subcommunities of CS with in parts very different belief systems, practices, tools, or traditions and therefore different cultures of learning and practices of enculturation. These can be for example: web development in the field of e-commerce, the development of embedded aviation systems, or distributed development of operating systems in open source projects. The relation to these different CoPs becomes particularly challenging when creating meaningful learning situations and teaching students programming practices beyond presentations of factual knowledge of programming syntax and exemplifying simple algorithms. What seems to be complex regarding tertiary CS education is even more so in secondary CS education. For instance, in Germany the primary goal of secondary CS Ed is not to engage and prepare students for tertiary CS Ed, but rather to enable students to interact consciously and well informed with information technology and use modeling skills "for understanding and solving problems in other fields of inquiry but also as tools for exploring and producing cultural artifacts" ([12], p. 6). Programming or coding is explicitly not regarded to be the most important practice students are supposed to adopt in CS class. From the situative perspective, this notion of CS Ed can be questioned as follows: What is the related CoP of these taught practices? What kinds of practices and tools determine the activities of this community and are they incorporated in the educational standards? What is the culture of learning and what are the practices of enculturation in CS class? What kind of pedagogy supports this enculturation and how does it need to be implemented regarding CS teacher education and lesson design? However, all these questions just refer to the notion of secondary CS Ed envisioned by the CS Ed community.

Student engagement is yet another facet to be taken into account: What kind of enculturation did students encounter before taking a CS class in high school? What are their expectations and belief systems regarding information technology and CS and is this aligned with the notion of CS Ed we are offering in secondary schooling? How is offered CS Ed contributing to their evolving identity as learners? Using a theoretical framework based on situated learning, Kolikant and Ben-Ari (2008) investigated how students and their teachers in a concurrent and distributed computing course in high school were interacting with each other. They observed different cultures of students and teachers that lead to a "clash of culture" and in consequence to learning difficulties in the CS classroom [16], see also ([26], p. 16). In order to overcome the cultural clash, Kolikant and Ben-Ari created a *fertile zone of cultural encounter*, a pedagogical innovation that bridges between student beliefs and expectations of CS and the language and formalisms of the professional culture presented by the CS teacher. This study exemplifies how the situated learning approach leads to questions of educational research and pedagogy beyond knowledge acquisition.

References

1. Anderson, J.R., Reder, L.M., Simon, H.A.: Situative versus cognitive perspectives: Form versus substance. Educational Researcher 26(1), 18–21 (1997)
2. Ben-Ari, M.: Situated learning in computer science education. Computer Science Education 14(2), 85–100 (2004)
3. Brown, J.S., Collins, A., Duguid, P.: Situated Cognition and the Culture of Learning. Educational Researcher 18(1), 32–42 (1989)
4. Chesñevar, C.I., González, M.P., Maguitman, A.G.: Didactic strategies for promoting significant learning in formal languages and automata theory. In: Proceedings of the 9th Annual SIGCSE Conference on Innovation and Technology in Computer Science Education (ITiCSE 2004), pp. 7–11. ACM (2004)
5. Collins, A., Brown, J.S., Holum, A.: Cognitive apprenticeship: Making thinking visible. American Educator 6(11), 38–46 (1991)
6. Daniels, M., Pears, A.: Models and methods for computing education research. In: Proc. Australasian Computing Education Conference (ACE). CRPIT, vol. 123, pp. 95–102 (2012)
7. Gramm, A., Hornung, M., Witten, H.: Email for you (only?): design and implementation of a context-based learning process on internetworking and cryptography. In: Proceedings of the 7th Workshop in Primary and Secondary Computing Education (WiPSCE), pp. 116–124. ACM, New York (2012)
8. Greeno, J.G.: On claims that answer the wrong questions. Educational Researcher 26(1), 5–17 (1997)
9. Hopcroft, J.E., Motwani, R., Ullman, J.D.: Introduction to Automata Theory, Languages, and Computation, 3rd edn. Pearson (2006)
10. Hundhausen, C.D.: Toward Effective Algorithm Visualization Artifacts: Designing for Participation and Communication in an Undergraduate Algorithms Course. Doctoral Thesis, University of Oregon, US (1999)

11. Kaptelinin, V., Nardi, B.: Acting with Technology –Activity Theory and Interaction Design. MIT Press, Cambridge (2004)
12. Knobelsdorf, M., Magenheim, J., Brinda, T., Engbring, D., Humbert, L., Pasternak, A., Schroeder, U., Thomas, M., Vahrenhold, J.: Computer Science Education in North-Rhine Westphalia, Germany – A Case Study. ACM Transactions on Computing Education 15(2) (2015)
13. Knobelsdorf, M., Tenenberg, J.: The context-based approach iniK in light of situated and constructive learning theories. In: Diethelm, I., Mittermeir, R.T. (eds.) ISSEP 2013. LNCS, vol. 7780, pp. 103–114. Springer, Heidelberg (2013)
14. Knobelsdorf, M., Isohanni, E., Tenenberg, J.: The reasons might be different – why students and teachers do not use visualization tools. In: Proceedings of the 12th Annual Finnish/Baltic Sea Conference on Computer Science Education (Koli), pp. 1–10. ACM, New York (2012)
15. Knobelsdorf, M., Kreitz, C., Böhne, S.: Teaching theoretical computer science using a cognitive apprenticeship approach. In: Proceedings of the 45th ACM Technical Symposium on Computer Science Education (SIGCSE), pp. 67–72. ACM, New York (2014)
16. Kolikant, Y.B.-D., Ben-Ari, M.: Fertile Zones of Cultural Encounter in Computer Science Education. Journal of the Learning Sciences 17(1), 1–32 (2008)
17. Lave, J., Wenger, E.: Situated Learning: Legitimate Peripheral Participation. Cambridge University Press (1991)
18. Malmi, L., Sheard, J., Simon, B.R., Helminen, J., Kinnunen, P., Korhonen, A., Myller, N., Sorva, J., Taherkhani, A.: Theoretical underpinnings of computing education research: what is the evidence? In: Proceedings of the International Workshop of Computing Education Research (ICER), pp. 27–34. ACM, New York (2014)
19. Nentwig, P.M., Demuth, R., Parchmann, I., Gräsel, I., Ralle, B.: Chemie im Kontext: Situated Learning in Relevant Contexts while Systematically Developing Basic Chemical Concepts. Journal of Chemical Education 84(9), 1439–1444 (2007)
20. Nunes, T.: What organizes our problem solving activities? In: Resnick, L., Saljo, R., Pontecorvo, C., Burge, B. (eds.) Discourse, Tools, and Reasoning: Essays in Situated Cognition, pp. 288–311. Springer (1997)
21. Pea, R.: Practices of distributed intelligence and designs for education. In: Salomon, G. (ed.) Distributed Cognition: Psychological and Educational Considerations, pp. 47–87. Cambridge University Press (1993)
22. Peters, A.-K., Rick, D.: Identity development in computing education: theoretical perspectives and an implementation in the classroom. In: Proceedings of the 9th Workshop in Primary and Secondary Computing Education (WiPSCE), pp. 70–79. ACM, New York (2014)
23. Rogoff, B.: The Cultural Nature of Human Development. Oxford University Press (2003)
24. Säljö, R.: Learning as the use of tools: a sociocultural perspective on the human-technology link. In: Littleton, K., Light, P. (eds.) Learning with Computers, pp. 144–161. Routledge, New York (1998)
25. Sfard, A.: On Two Metaphors for Learning and the Dangers of Choosing Just One. Educational Researcher 27(2), 4–13 (1998)
26. Tenenberg, J., Knobelsdorf, M.: Out of Our Minds: A Review of Sociocultural Cognition Theory. Computer Science Education 24(1), 1–24 (2014)
27. Vygotsky, L.S.: Mind in Society: The Development of Higher Psychological Processes. Harvard University Press (1978)
28. Vygotsky, L.S.: The instrumental method in psychology. In: Wertsch, J.V., Sharpe, M.E. (eds.) The Concept of Activity in Soviet Psychology (1981)

Robotics Activities–Is the Investment Worthwhile?

Ronit Ben-Bassat Levy and Mordechai (Moti) Ben-Ari

Department of Science Teaching
Weizmann Insitute of Science
Rehovot 76100 Israel
{ronit.ben-bassat,moti.ben-ari}@weizmann.ac.il
http://www.weizmann.ac.il/sci-tea/benari/

Abstract. Young people are deterred from studying *science, technology, engineering and mathematics (STEM)* by the perception that such studies are boring and by a lack of self-efficacy. One approach towards increasing engagement with STEM is through the use of robotics in education, both in formal instruction and through informal activities such as competitions. There is a consensus that such activities are "fun" but there is almost no research on whether there is any educational advantage to robotics activities. We are investigating the extent to which participation in robotics education activities influence the attitudes of students towards STEM and their intentions concerning STEM studies in the future. The research framework and methodology is the *theory of planned behavior (TPB)*, which claims that *attitudes* engender *intentions*, which in turn cause *behavior*. TPB is based upon questionnaires that are constructed based upon observations and interviews. The analysis of the answers from 106 questionnaires showed that the attitudes and the subjective norms were not as high as we expected, but the results for the subjective norms are of particular importance, because they show that students can be motivated by the respect and support they receive from their teachers and parents. The scores for the intentions predictor were very high, which implies that the students are like to choose to study STEM in the future.

Keywords: robotics, theory of planned behavior.

1   Introduction

Many factors discourage students from studying *science, technology, engineering, mathematics (STEM)*. Some believe, incorrectly, that most jobs are being outsourced [9]. More importantly, working in STEM is perceived as boring and monotonous, only appropriate for nerds [11]. This perception functions as a serious deterrent to female students [17].

Attitudes concerning STEM are formed as early as middle school [13]; therefore, if one hopes to influence students attitudes it must be done early. This is a

© Springer International Publishing Switzerland 2015
A. Brodnik and J. Vahrenhold (Eds.): ISSEP 2015, LNCS 9378, pp. 22–31, 2015.
DOI: 10.1007/978-3-319-25396-1_3

justification for teaching subjects such as computer science (CS) to high-school students and even to middle-school students. One approach is to use kinesthetic activities, for example, *Computer Science Unplugged* (http://csunplugged.org/). Another is to use programming environments designed for young students [15]: Scratch (http://scratch.mit.edu/) and Alice (http://www.alice.org/).

There is a third approach that has become very popular: teaching with robotics. This became feasible with the appearance of LEGO Mindstorms in 1998. The FIRST LEGO League (FLL) conducts worldwide robotics challenges with 16762 teams participating in 2010 (http://www.firstlegoleague.org/).

Recent advances have made robotics even more accessible. The *Institute for Personal Robots in Education* at the Georgia Institute of Technology and Bryn Mawr College (http://roboteducation.org/) combined an off-the-shelf robot (the *Scribbler*) with a circuit board that added sensors and Bluetooth communication. There is a textbook that accompanies the robot and its software [16]. Similar educational robots are the *Finch* developed at Carnegie Mellon University (http://www.finchrobot.com/) and the *Thymio-II* developed at the École Polytechnique Fédérale de Lausanne (https://www.thymio.org/). Compared with kinesthetic activities and visual programming environments, robotics has the additional advantage that students can learn a variety of STEM subjects in this context, not just computing.

Young people enjoy working with robots and they are proud of their achievements. The question we are asking is whether the result of engaging in robotics transcends fun and leads to significant positive changes in their attitudes towards STEM and in their intentions to continue studying STEM. It is essential to answer this question because of the massive investment of time, money and effort in robotics, an investment that can be justified only if research shows that the above goals are achieved.

2   Background

2.1   Research on Young Students Learning CS

Research on attracting young students to CS has been performed using the Alice visual programming environment. Adams described a CS summer camp for middle-school students that increased their willingness to engage with computing technology [3]. Similar results were reported by Kelleher [14].

Taub, Ben-Ari & Armoni [21] found that—although students enjoy the CS Unplugged activities—there was little effect on their attitudes towards CS nor on their intentions to pursue CS. Meerbaum-Salant, Armoni & Ben-Ari [20], investigated the Scratch visual programming environment. They showed that middle-school students can successfully learn concepts of CS, but that the learning is sub-optimal unless the teacher is knowledgeable in the subject matter and actively engages in guiding the students. In subsequent research [8], they showed that middle-school students who learned Scratch found it easy to learn professional programming languages (Java and C#) in high school.

2.2   Robotics

The literature on learning through robotics is quite extensive. A few representative collections are: the double issue of the *ACM Journal of Educational Resources in Computing* on robotics [1,2] and the book edited by Druin & Hendler [12]. Unfortunately, most of this work is anecdotal.

Verner & Ahlgren [22] held robotics contests and examined the content knowledge of the students following the contests. They found that students at all levels (middle school, high school and university) obtained relatively low scores on a test that evaluated their knowledge of the various subject areas relevant to robotics. Nevertheless, students reported progress in understanding content and in improving learning skills.

Anderson et al. [6] used robotics labs to augment a first-year university course in CS. They were particularly interested in lowering the intimidation felt by novices in CS courses and in increasing interest in majoring in the subject. The results of their surveys were mixed and only partially significant, with some students showing increased interest and lower intimidation, while others displayed decreased interest and higher intimidation.

Three recent research projects investigated students' motivation to learn CS in the context of robotics activities.

Markham & King [18] taught a CS1 course with one section using the Scribbler robot and the other a control group. Using surveys, they found that the robotics group devoted more effort when compared with non-robotics classes; they claimed that this extra effort implies increased intrinsic motivation.

Apiola et al. [7] used LEGO Mindstorms in a voluntary university course. The researchers interviewed students and determined that they found working with robots fun and exciting, but also challenging and frustrating. However, the lack of a traditional support structure proved detrimental to the students' achievements.

The study by McGill [19] on increasing motivation through the use of robotics is more relevant because it was done with students in a CS0 course who did not intend to specialize in CS. She found that the use of the Scribbler robots improved students attitudes towards programming, but had little effect on other measures such as confidence.

2.3   The Research in Light of Previous Work

Our research significantly extends existing work in several aspects:

- We investigate attitudes towards STEM in general and not just towards CS or robotics.
- Our research methodology goes beyond measuring attitudes and looks into the intentions that are engendered by the attitudes. This is important because it is intentions that directly affect future behavior.
- By carrying out the research in middle schools, we are checking the effect of robotics at critical ages where students may not have made firm decisions as to their future.

3   Description of the Research

3.1   Research Question

To what extent does participation in robotics activities influence the attitudes of students towards STEM and their intentions concerning STEM studies in the future?

We conjecture that participation in robotics activities will positively influence students' attitudes and their intention to study STEM. If this conjecture is verified, it would provide a research-based justification for the extensive investment in robotics in education.

3.2   Population

The research is being carried out in the context of two robotics activities. The first population is middle-school students participating in the FIRST LEGO League competition. These activities are extracurricular and the participants are self-selected. The second population is middle-school students participating in a new program of the Ministry of Education [23]. This program is part of the school curriculum and the participants are selected by the teachers and principals. Therefore, these students are likely to display a more diverse set of attitudes and intentions.

3.3   Research Framework

The research is based on the *theory of planned behavior (TPB)* [5]. This is both a theoretical research framework from social psychology and a quantitative methodology. TPB predicts that behavior results from intentions, which in turn are formed from attitudes towards the behavior, in this case, choosing to study STEM subjects in the future. TPB claims that three variables—attitudes, subjective norms and perceived behavior control—predict the intention to perform a behavior. We now give definitions of the *predictors* that appear in the model:

Behavior Behavior is an observed human action that is a response to a given situation. In TPB, behavior is a function of intentions and perceptions of behavioral control.

Intention Intention is an indication of a person's readiness to perform a given behavior and is considered to be the immediate antecedent of behavior. TPB does not predict behavior only according to intentions; these are combined with attitudes, subjective norms, and perceived behavior control.

Attitude towards a behavior Attitude towards a behavior is the degree to which the performance of the behavior is positively or negatively valued. This evaluation of the behavior is assumed to have two components which work together: (1) beliefs about the consequences of the behavior (behavioral beliefs), and (2) the corresponding positive or negative judgments about each of these consequences (outcome evaluations).

Subjective norms about the behavior Subjective norms are a person's estimate of the social pressure to perform or not to perform the target behavior. The subjective norms contain two connected elements: (1) the beliefs of other people that may be important to the person and how those other people want the person to behave (normative beliefs), and (2) the positive or negative evaluation of each belief (motivation to comply).

Perceived behavioral control Perceived behavioral control is the extent in which a person feels that he can control the behavior. Perceived behavior control has two aspects: (1) how much control does a person have on the behavior (control beliefs), and (2) how confident a person feels about his ability to behave in a certain way (influence of control beliefs).

We have used TPB within the context of educational technology [10] and it proved very effective in understanding the causal links from attitudes to intentions to behavior. TPB is particularly suited to this research project because it is capable of uncovering fine distinctions and interesting phenomena that might be overlooked by other methods.

3.4   Research Instruments and Data Analysis

The first year of the research program was devoted to field observations and interviews. Data from these observations and interviews were analyzed to discover issues that related to the different TPB predictors. We used the approach in [4] to guide the formulation of questions regarding goal-directed behavior, focusing on behavioral beliefs, normative beliefs, control beliefs and intentions. In addition, questions were formulated to investigate the strengths of the outcomes of each of these constructs.

There were 44 questions that used a seven-point Likert scale (from "strongly disagree" to "strongly agree"). Multiple questions were used to ensure reliability. The association of a question with a construct was performed by the first researcher and then validated by the second researcher; disagreements were discussed until a consensus was reached. We sent out more than 700 questionnaires and have so far received the answers from about 350, of which we analyzed 106. The large number of responses should lead to significant results.

The questionnaire can be viewed at `http://goo.gl/forms/xu8NuDLLtI`.

The analysis was performed according to the methods of TPB. For each of the predictors—attitudes, perceived behavior control, intentions—the analysis proceeded through the following steps: (a) the data were sorted in ascending order and divided into quartiles; (b) each subject was ranked into one of the quartiles according to his/her answers; (c) connections were sought among the rankings; (d) conclusions about the students' attitudes and intentions were then inferred from these rankings.

3.5 The Questionnaire

Here are some of the questions together with an explanation of their design:

Question 8 *I believe that the process of trial-and-error (which I experience during robotics classes) makes me feel like a real scientist.* This question originated from students who expressed statements like "I am a scientist," "I am trying to solve the problem using the robot," "I can make mistakes and learn from them." The question investigates behavioral belief.

Question 11 *To be involved in science in the future will mean success for me.* A student who gives a high score on this question strengthens the significance of the answer to question 8, and, conversely, a low score reduces the significance of the answer.

Question 30 *It is difficult for me to study in the afternoons, because I have to take care of my brothers and sisters.* This question originated from complaints of students who could not continue participating in the robotics activities because of commitments such as homework, other extracurricular activities, and taking care of siblings. The question investigates the students' control beliefs.

Question 15 *Studying in the afternoons is difficult for me, so I don't want to learn robotics.* This question checks the power of the control beliefs investigated by question 30.

Question 22 *The teachers at the school respect me more because I study robotics.* This question originated from students who remarked that they are better respected since they began to study robotics. The question investigates normative beliefs.

Question 28 *It is important to me what the teachers think of me.* This question measures the normative belief motivation to comply checked in question 22.

4 Results

From the observations and the interviews we found that most of the students were enthusiastic towards robotics at the beginning of the year when the subject was new. They carried out the assignments given by teachers and they collaborated on the construction of the robot. The FLL students collaborated more than the students in the classroom because they had a concrete goal, while the other students became bored. The robots in several classes did not function as expected, which led to frustration. The interviews showed that the students felt good when they received respect and support from the teachers and the staff at the school, as well as from their parents. The interviews revealed a problem in scheduling: robotics classes are usually given in the afternoon after all the other students have gone home.

Each of the predictors was calculated according to the TPB framework: the sum over all relevant questions of multiplication of the score of the answer and its associated strength:

Attitudes   = Behavioral beliefs × Outcome evaluation
Subjective norms   = Normative beliefs× Motivation to comply
Perceived behavioral
   control (PBC)   = Control beliefs × Power of control belief

For example, for the attitudes, question 1 measured behavioral beliefs and question 11 measured its outcome evaluation; therefore, the score for question 1 was multiplied by the score for question 11. Similarly, this was done for other pairs of questions and the results added to obtain the measure of the attitudes.

We calculated the minimum and maximum *possible* scores, as well as the average score calculated over all the 106 students:

	Attitudes	Subjective norms	PBC	Intentions
Minimum score	9	9	12	3
Average score	243	241	184	17
Maximum score	441	441	294	21

The average score is roughly halfway between the minimum and maximum scores. This is a disappointing result since we hoped to find much higher scores, indicating high positive attitudes. On the other hand the results show that the students do not possess negative attitudes, even though such attitudes were sometimes expressed in the preliminary interviews.

In order to obtain a more fine-grained presentation of the results, we divided the results into four quartiles and placed the students in those quartiles. The advantage of quartiles over the average is that it enables a deeper analysis of the each of the predictors.[1]

The following table shows the quartiles of each predictor:

	Attitudes	Subjective norms	PBC	Intentions
First quartile	342-441	342-441	225-294	18-21
Second quartile	242-341	242-341	154-224	13-17
Third quartile	142-241	142-241	84-153	8-12
Fourth quartile	9-141	9-141	11-82	3-7

---

[1] The division into quartiles is not an aspect of TPB as found in the literature; we first used this technique in [10].

and this table shows the number of students in each quartile:

	Attitudes	Subjective norms	PBC	Intentions
First quartile	27	12	34	58
Second quartile	24	36	33	32
Third quartile	31	53	37	12
Fourth quartile	24	5	2	4
Total	106	106	106	106

- The students are roughly uniformly distributed in the quartiles for attitudes, which is somewhat disappointing.
- Most of the students fall into the two middle quartiles for subjective norms, which means that they can be influenced to choose STEM by the school and home environments.
- The relatively high scores for perceived behavioral control mean that students feel that they can control their future choices to study STEM.
- The scores for intentions are very high, indicating that they are likely to choose STEM.

5 Discussion

The observations and interviews enabled us to construct a 44-item questionnaire according to the TPB framework, as demonstrated by the examples in Section 3.5. The observations and interviews revealed several aspects of students' engagement in robotics activities (Section 4). While the attitudes and subjective norms were not as high as we expected, the results for the subjective norms are of particular importance, because they show that students can be motivated by the respect and support they receive from their teachers and parents.

The high scores for intentions were compatible with a direct question asked in personal questions that appeared before the TPB part of the questionnaire:

Do you intend to study a scientific subject (biology, chemistry, physics, computer science) in high school? Yes / No

This strengthens our belief that robotics encourages intentions to choose STEM.

6 Conclusions

The theory of planned behavior proved to be a fruitful framework for research into students' engagement in robotics activities. We believe that a questionnaire constructed according TPB and based upon field work is a better instrument than a questionnaire based only upon the researchers' intuition and experience.

Robotics activities are often justified on the claim that they motivate students to pursue further studies in STEM subjects. Our results provide empirical evidence that supports this claim.

References

1. ACM: Journal of Educational Resources in Computing 3(4) (2003)
2. ACM: Journal of Educational Resources in Computing 4(3) (2004)
3. Adams, J.C.: Alice, middle schoolers & the imaginary worlds camps. SIGCSE Bulletin 39(1), 307–311 (2007)
4. Ajzen, I., Madden, T.: Prediction of goal-directed behavior: Attitudes, intentions, and perceived behavioral control. Journal of Experimental Social Psychology 22, 453–474 (1986)
5. Ajzen, I.: Perceived behavioral control, self-efficacy, locus of control, and the theory of planned behavior. Journal of Applied Social Psychology 32(4), 665–683 (2002)
6. Anderson, M., McKenzie, A., Wellman, B., Brown, M., Vrbsky, S.: Affecting attitudes in first-year computer science using syntax free robotics programming. ACM Inroads 2(3), 51–57 (2006)
7. Apiola, M., Lattu, M., Pasanen, T.: Creativity and intrinsic motivation in computer science education: experimenting with robots. In: 15th Annual Conference on Innovation and Technology in Computer Science Education (ITiCSE 2010), pp. 199–203 (2010)
8. Armoni, M., Meerbaum-Salant, O., Ben-Ari, M.: From Scratch to "real" programming. ACM Transactions on Computing Education 14(4), article 25 (2015)
9. Aspray, W., Mayadas, F., Vardi, M.Y. (eds.): Globalization and Offshoring of Software: A Report of the ACM Job Migration Task Force. ACM (2006), http://www.acm.org/globalizationreport/ (last accessed on March 10, 2015)
10. Ben-Bassat, L.R., Ben-Ari, M.: Adapting and merging methodologies in doctoral research. Computer Science Education 19(2), 51–67 (2009)
11. Carter, L.: Why students with an apparent aptitude for computer science don't choose to major in computer science. SIGCSE Bulletin 38(1), 27–31 (2006)
12. Druin, A., Hendler, J. (eds.): Robots for Kids: Exploring New Technologies for Learning. Morgan Kaufmann (2000)
13. Gibbons, S.J., Hirsh, L.S., Kimmel, H., Rockland, R., Bloom, J.: Middle school students attitudes to knowledge about engineering. In: International Conference on Engineering Education, pp. 1–6 (2004)
14. Kelleher, C.: Motivating Programming: Using Storytelling to Make Computer Programming Attractive to More Middle School Girls. Ph.D. dissertation, Carnegie Mellon University (2006), http://www.cs.cmu.edu/~caitlin/kelleherThesis_CSD.pdf (last accessed on March 10, 2015)
15. Kelleher, C., Pausch, R.: Lowering the barriers to programming: A taxonomy of programming environments and languages for novice programmers. ACM Computing Surveys 37(2), 83–137 (2005)
16. Kumar, D.: Learning Computing with Robotics (2011), http://cs.brynmawr.edu/~dkumar/Myro/Text/Fall2011PS2/PDF/LCR2011.pdf (last accessed on March 10, 2015)
17. Margolis, J., Fisher, A.: Unlocking the Clubhouse: Women in Computing. MIT Press (2003)
18. Markham, S., King, K.: Experiences, outcomes, and attitudinal influences. In: 15th Annual Conference on Innovation and Technology in Computer Science Education (ITiCSE 2010), pp. 204–208 (2010)
19. McGill, M.: Learning to program with personal robots: Influences on student motivation. ACM Transactions on Computing Education 12(1), article 4 (2012)

20. Meerbaum-Salant, O., Armoni, M., Ben-Ari, M.: Learning computer science concepts with Scratch. Computer Science Education 23(3), 239–264 (2013)
21. Taub, R., Ben-Ari, M., Armoni, M.: The effect of CS Unplugged on middle-school students views of CS. In: 14th Annual ACM SIGCSE Conference on Innovation and Technology in Computer Science Education (ITiCSE 2009), pp. 99–103 (2009)
22. Verner, I., Ahlgren, D.: Robot contest as a laboratory for experiential engineering education. ACM Journal on Educational Resources in Computing 4(2), 2–28 (2004)
23. Zur Barguri, I.: A new curriculum for junior-high in computer science. In: 17th Annual Conference on Innovation and Technology in Computer Science Education (ITiCSE 2012), pp. 204–208 (2012)

Dimensions of Programming Knowledge

Andreas Mühling, Peter Hubwieser, and Marc Berges

TUM School of Education
TU München
{andreas.muehling,peter.hubwieser,marc.berges}@tum.de

Abstract. Nowadays, learning and teaching outcomes are defined predominantly by target competencies. In order to assess learning outcomes, properly defined and empirically validated competency models are required. For object-oriented programming, such models have not been brought forward up to now. Aiming to develop a competency structure and level model for this field, we have examined the structural knowledge of programming novices to derive its potential dimensions. The results suggest 6 dimensions. Additionally, we propose difficulty levels for two of these dimensions based on the SOLO taxonomy. The empirical validation of these dimensions and their levels is subject to further investigations.

1 Introduction

Traditionally, the desired learning outcomes of schools were defined and controlled by the acquisition of certain knowledge elements or by the achievement of predefined learning objectives. During the last decade, mainly stimulated by the surprising results of the PISA (Programme for International Student Assessment) studies [24], the focus of the outcomes of school education was shifted more and more towards target competencies in many countries. In this context, the term competency is usually understood in the sense of Weinert, who defined it as "the cognitive abilities and skills possessed by or able to be learned by individuals that enable them to solve particular problems, as well as the motivational, volitional and social readiness and capacity to use the solutions successfully and responsibly in variable situations"[36].

To align learning and teaching processes and measure their success, these target competencies have to be defined and structured properly by suitable empirically validated competency models. For this purpose, different kinds of models are used [13], which may focus on the structure, the different hierarchical levels or the development of the relevant competencies [14]. Regarding the definition and measurement of competencies, much groundbreaking work was done in the context of the PISA studies (see e.g. [31], [24]).

For Computer Science Education, the development process of competency models is just at its beginning. As far as we know, only one attempt has been made until now that could cope with the standards of PISA [15], [18]. Yet, as the scope of this project is very broad, it does not intend to provide sufficiently detailed models of programming competencies. Thus our team has set its research focus on the development of detailed competency structure models for

© Springer International Publishing Switzerland 2015
A. Brodnik and J. Vahrenhold (Eds.): ISSEP 2015, LNCS 9378, pp. 32–44, 2015.
DOI: 10.1007/978-3-319-25396-1_4

programming. To keep things simple, we decided to start these investigations in the field of object-oriented programming and restricted to programming novices.

Regardless the complex structure of competencies according to Weinert [36], they remain "cognitive abilities and skills". Thus the development of competency models has to start with the investigation of the relevant cognitive structures [14], providing a skeleton. The position of a competency in structure models is then defined by the location of its corresponding knowledge element. Based on large sets of concept maps that had been gathered by our team over several years in different programming courses, 6 potential dimensions of programming knowledge were identified at the end [17]. Aiming to validate these knowledge structures and find competency definitions, we are conducting qualitative investigations of assignments, tasks (see [27]), structured interviews and learning diaries.

In this paper, we present the first results and discuss the relevance for our intended competency model. In the next steps, we are going to construct and validate items for the further investigation of the 6 dimensions. The outcomes of the other activities will be presented in subsequent publications.

2   Background and Related Work

Basically, we aim to combine two different aspects of knowledge respectively competencies in our model: structure and difficulty. To investigate the first aspect, we have analyzed large sets of concept maps. For the second, we propose a scale that is inspired by the SOLO taxonomy of Biggs.

2.1   Concept Maps and Declarative Knowledge

Concept maps were invented in the 1970s as a tool to help structuring and visualizing the responses of children in clinical interviews [20]. Later on, the use of concept maps shifted from a specific technique for data analysis to a general technique for learning, teaching, and assessing structural knowledge [21].

A concept map consists of labeled entities that represent a *concept*. Concepts are defined as "perceived regularities or patterns in events or objects, or records of events or objects, designated by a label" [20]. Two concepts that are linked by a connection are forming the basis of a *proposition*. A proposition is composed of the two concepts and the label of the connection itself. The technique of concept mapping is fundamentally based on the ideas of Constructivism and meaningful learning [23].

The subject-matter knowledge about programming resides in a part of the long-term memory, called *declarative memory* [19], that can be described rather succinctly: "Declarative memory is memory that is directly accessible to conscious recollection. It can be declared. It deals with the facts and data that are acquired through learning" [34]. In consequence, a concept map can be regarded as an externalization of declarative knowledge as it has been "declared" by the concept mapper [28]. In general, the organization of concepts in memory is assumed to be pivotal for the quality of a person's knowledge [35]. Without a structured connection to others, a concept will not be kept in long term memory [33]. According to [28], "concept interrelatedness is an essential property of knowledge'".

Literature points to the successful application of concept maps as a method for assessing or evaluating students' knowledge structures across many different fields of study [22], [20]. In particular, it has been shown that assessments based on concept mapping can differentiate between the knowledge of experts and novices as well as between meaningful learning and rote-learning [7]. Concerning the validity of concept map assessment tasks, the results found in literature are generally positive [22], [26]. The reliability has been established, at least under certain conditions, by several different studies [22]. In educational research about computer science, concept maps have been used extensively, see e.g. [8], [12], [29].

2.2   The SOLO Taxonomy

In 1982, Biggs proposed his SOLO taxonomy, see Table 1. Based on the theory of meaningful learning [4], it puts more emphasis on the learner and the actual *learning outcome*, instead of the learning material [4]. In Table 1, Capacity "refers to the amount of working memory, or attention span, that the different levels of SOLO require" [4]. The relating operation "refers to the way in which the cue and response interrelate" [4]. Additionally, there is an attribute of "Consistency and closure", referring to the felt need of the learner to come to a conclusion that is consistent with the data and other possible conclusions [4], which increases with the levels of the taxonomy.

Table 1. The levels of the SOLO taxonomy as described by [4]

SOLO Level	Capacity	Relating operation
Prestructural	Minimal: cue and response confused	Denial, tautology, transduction. Bound to specifics.
Unistructural	Low: cue and one relevant datum	Can "generalize" only in terms of one aspect.
Multistructural	Medium: cue and isolated relevant data	Can "generalize" only in terms of a few limited and independent aspects.
Relational	High: cue and relevant data and interrelations	Induction: Can generalize within given or experienced context using related aspects.
Extended Abstract	Maximal: cue and relevant data and interrelations and hypotheses	Deduction and induction. Can generalize to situations not experienced.

Regarding the application of the SOLO taxonomy in programming education, Hawkins [10] proposed different programming patterns of novices that correspond to the categories of Biggs [4] for the exemplary task of drawing simple shapes as circles or rectangles. He associated

1. Pre-structural Response: Immediate mode, commands are applied by trial and error, until the result is acceptable,
2. Unistructural Response: Immediate mode, the commands are entered in a planned and deliberated sequence,
3. Multistructural Response: Programming mode, structured sequences, and
4. Relational Response: Functions are defined and Control Structures are use. Code is reused
5. Extended Abstract Response: Parametrized functions.

2.3   The Cognitive Structure of Programming

Since the first days of computer programming, scientists reason about the corresponding cognitive structures. At this place, we have to restrict ourselves to the most relevant sources for our project.

In 1990, Ormerod partitioned the knowledge Schema of "Programming Language Components" into Data Structures (Arrays, Lists) and Looping Constructs (Recursion, Iteration) [25].

Complexity in object-oriented programming can increase due to more complex methods or by adding classes and connections between them. Shao et al. introduced in [32] a metric for programming based on cognitive weights of basic control structures (BCS). "The cognitive weight of software is the degree of difficulty or relative time and effort required for comprehending a given piece of software modelled by a number of BCSs" The metric counts the basic control structures in source code. Misra et al. generalized the method to object-oriented programming defining the complexity as the addition of the complexity of the involved methods and objects [16].

In 2006 Bennedsen and Schulte introduced a theoretical base for measuring the understanding of object interaction [2], which was revised in 2013 [3], describing four hierarchical levels: Interaction with objects, Interaction on object structures and Interaction on dynamic object structures.

3   The Educational Context

In the year 2004, the government of the German state of Bavaria introduced a new compulsory subject of Computer Science (CS) in all its 400 Gymnasiums (grammar schools) [11]. This subject was introduced simultaneously with the reduction of the Bavarian Gymnasium from 9 to 8 grades. As a consequence, in April 2011 two different age cohorts graduated simultaneously from the Bavarian Gymnasium: the last age group of G9 and the first one from the new G8. The latter also represents the first cohort who had attended the new CS classes over 2-6 years. The goal of our study was to compare the declarative knowledge of these two groups. Regarding the learning content that is relevant for our study, the CS course is based on a "strictly objects first" approach. The courses start in grade

6 with the introduction of the concepts object, attribute, method and class in the context of vector graphics, followed by the concept of aggregation. Subsequently, the students work with recursive aggregation, applied to file trees. In the next step, the trees are generalized to graphs. At the end of grade 7, the students work out their first programs, using a virtual robot. In the 9th grade, the students learn to apply the concept of function, followed by (object-oriented) data modeling. The students of grade 10 consolidate their object-oriented knowledge by "real" object-oriented programming, designing object-, class- and state-models and implementing them with a suitable object-oriented programming language, which currently will be Java in most of the schools. Additionally, they learn to apply the concepts of sub- and superclass, inheritance and polymorphism. In the elective courses in grade 11, the object-oriented modeling and programming concepts are extended by the recursive data structures List, Tree and Graph. For more details about the this course see [11].

4   The Study

4.1   Design

We conducted a cross-sectional study among the beginning computer science students at our university in the summer and winter term of 2011. We were able to present our questionnaire to 590 of the about 700 freshmen. 338 of the responses contained relevant information. The analysis was restricted to the 290 concept maps of the graduates from the Bavarian G8 or G9 school system(see section above). For more details, see [17].

The participants were given a list of 40 concepts, which had been extracted from the curriculum of the school subject semi-automatically by extracting nouns, counting their frequency and then manually filtering the list. The remaining 40 concepts were: *algorithm, array, assignment, association, attribute, automaton, class, conditional statement, data, data structure, edges, flow, grammar, graph, instruction cycle, list, loop, method, object, processor, program, programming language, record, recursion, register machine, semantics, sequence, state, state change, statement, subclass, superclass, syntax, table, tree, pointer, variable, value, vertex, working memory.*

First, the participants were asked to mark all concepts they were familiar with and afterwards draw a concept map of those. Several different orderings for the concept list were used in order to measure whether this has an effect on the maps, but none was found. The students received a short written introduction on concept mapping including an example map from Mathematics. The participants were asked to label the propositions. The maps were produced with paper and pencil. For the whole survey, the students were given 45 minutes time.

4.2   Data Analysis

The basic idea of the analysis was to create concept landscapes from sets of maps [17]. These landscapes could then be analyzed in several ways. For this work, only one method was used:

1. Create a weighted graph from a set of concept maps - the edge weights reflect the number of maps in which a given pair of concepts has been connected by a proposition.
2. Use graph algorithms on this weighted graph - in our case community detection using a Greedy algorithm [5]
3. Prepare the graph for qualitative, visual analysis by pruning edges. To this end, the Pathfinder algorithm was used.

Pathfinder networks as described by Schvaneveldt are graph based representation of the similarity (or dissimilarity) of entities [30]. Originally, the data that is represented consists of pairwise similarity ratings given by persons, usually using a numeric scale. The similarity ratings can be modeled as a weighted, complete graph with each entity becoming a node and the weight of each edge is the similarity value of the pair of incident entities of that edge. Such a representation is called a "network" [30]. Schvaneveldt note that "[a]s psychological models, networks entail the assumption that concepts and their relations can be represented by a structure consisting of nodes (concepts) and links (relations). Strength of relations are reflected by link weights and the intentional meaning of a concept is determined by its connections to other concepts". Algorithmic methods can then be used to analyze such a network, or extract prominent features. The Pathfinder algorithm is one such method and an alternative is, for example, *multi-dimensional scaling* (MDS) developed by Kruskal [1].

As described by Dearholt & Schvaneveldt [6], a Pathfinder network is always constructed from an existing, non-negative weight matrix, which represents a weighted graph. The Pathfinder network is itself again a graph that is directed if and only if the input graph was directed. It consists of the same nodes and components as the input graph but of a subset of its edges, with their weights preserved. The edges are chosen such that the final network provides a path of minimal distance between each pair of nodes according to a special metric (called *Minkowski-* or *r-metric*) that is dependent on a parameter $r > 0$. The weight of a path consisting of edges $e_1, e_2, \cdots, e_k$ with weights $w_1, w_2, \cdots, w_k$ according to the r-metric is defined as:

$$(w_1^r + w_2^r + ... + w_k^r)^{1/r}$$

For $r = 1$ the r-metric defaults to the sum of the single edge weights, for $r = 2$ it is the Euclidean distance and for $r = \infty$ the path weight is the maximum edge weight along the path [6]. These three values are representing highly used metrics and are called *Manhattan distance, Euclidean distance*, and *Chebyshev distance* respectively [9].

Additionally, a Pathfinder network with n nodes is guaranteed to be *q-triangular* for $q \in \{1, 2, ..., n - 1\}$. This means, that the weight of any edge (i, j) is less or equal than the weights, according to the chosen r-metric, of any path between i and j that is of length at most q [6]. In other words, when ignoring paths longer than q, the triangle inequality holds for each pair of nodes in the graph. If q is set to the maximal value of $n - 1$, the (regular) triangle inequality always holds.

For our analysis, the maps were split into two groups according to the students' prior education. *G8* denotes the group of students who attended the compulsory subject Informatics for at least four years and *G9* denotes the group who didn't have a compulsory subject. The G8 group consists of 163 maps and the G9 group of 127.

The first result was that G8 maps were denser. As a t-test of the number of edges between two groups showed, the hypothesis "the true difference in means is 0" could be rejected with a confidence level of 99% ($p = 0.0001$).

To identify the prevalent structures, both sets of maps were transformed into a concept landscapes and the Pathfinder networks with parameters $q = 39$ and $r = \infty$ were created for each landscape separately. Edges that were appearing in less than 10% of the maps of each group were removed beforehand and unconnected concepts were also removed. First, taking a look at the concepts that remained in the Pathfinder networks, there were 31 for G8 and 26 for G9. The networks are shown in Figure 1. All concepts appearing in the G9 networks are also appearing in the G8 network, however, the G8 network contains the additional concepts: *list, recursion, semantics, state,* and *register machine.*

Analyzing the communities in the Pathfinder networks using a a greedy algorithm, both networks are partitioned into six communities. The concepts are assigned to the communities as shown in Table 2.

Table 2. The communities identified within the Pathfinder network from the maps of the G8 and G9 groups. The ordering of the communities for both groups is arbitrary, but was chosen to allow comparing the groups more easily.

Community	G8	G9
1	*Data structure, Graph, Edges, Nodes, List, Tree, Array, Recursion*	*Data structure, Graph, Edges, Nodes, Tree, Array*
2	*Data, Record, Table, Working memory, Processor, Register machine*	*Data, Record, Table, Working memory, Processor*
3	*Object, Attribute, Value, Variable*	*Object, Attribute, Method, Algorithm*
4	*Class, Method, Loop, Conditional Statement*	*Class, Value, Variable, Loop, Conditional Statement*
5	*Programming language, Syntax, Semantics, Grammar*	*Programming language, Syntax, Grammar*
6	*Statement, Program, Automaton, State, Algorithm*	*Statement, Program, Automaton*

4.3   Discussion

Taking together the Pathfinder networks of Figure 1 and the communities of Table 2, the following differences can be observed.

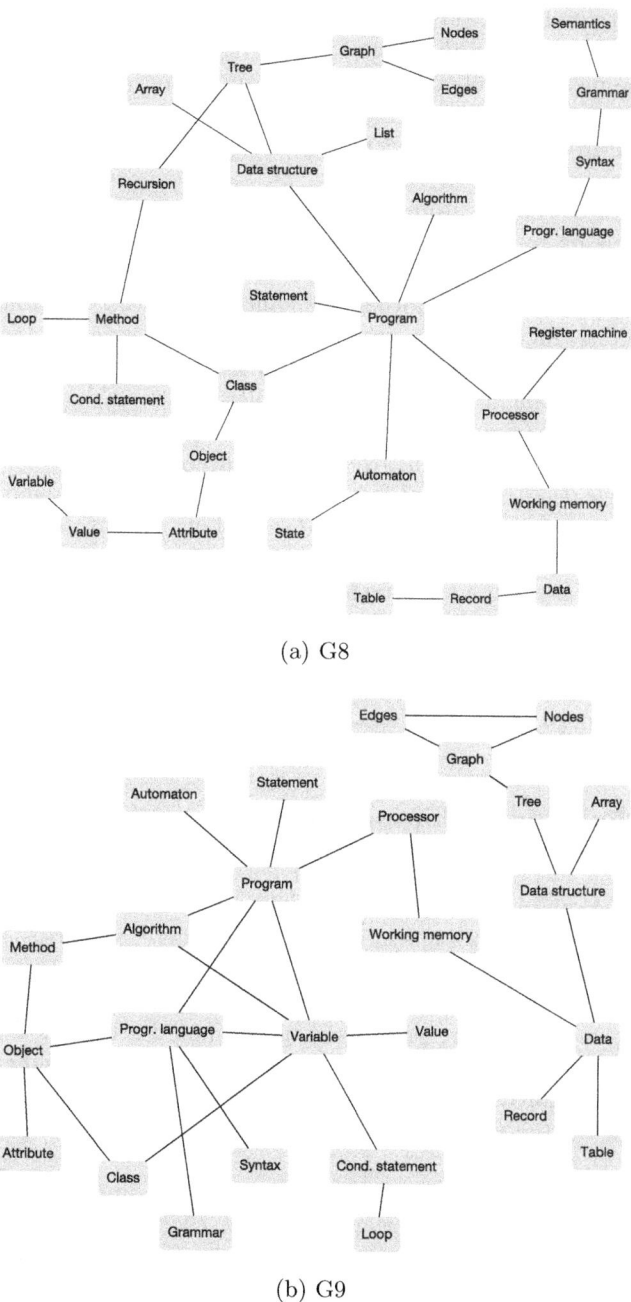

(a) G8

(b) G9

Fig. 1. The Pathfinder networks of students with compulsory school subject and of students without.

First of all, the G8 network is visibly more complex, because the maps of the G8 groups are significantly denser. Furthermore, while it seems common among all beginning students to value the concept *program* highly in their knowledge structure, the most central concepts of *class* and *data structure* as opposed to *processor* and *data* show a more object-oriented understanding of the G8 group, while the G9 group seems to be more focused on the technical aspects and computers themselves.

The network of the G8 group has a connection between *recursion* and *tree*, which corresponds to the approach of introducing recursion based on object-oriented recursive data structures, like lists and trees, chosen in the curriculum. Another indicator for a more formal CS education in the G8 group: For their network, there is the path *programming language, syntax, grammar, semantics,* whereas the G9 group only has *grammar* and *syntax* connected to *programming language* directly (*semantics* is missing completely).

There are some commonalities between the two groups: First, communities 1, 2, and 5 are nearly identical except for concepts missing the G9 network altogether. Next, there is a clear grouping of concepts related to data structures for both groups in the first community. Also, for both groups, the more database oriented concepts (*data, record, table*) are grouped with *processor* and *working memory*, placing them in a more technical and less abstract corner. In the same vein, *register machine* is seemingly more related to a real processor than to an abstract notion for beginning students. Finally, for both groups, *automaton, program* and *state* are related; however, none of the object-oriented concepts are put into that group, indicating a lack of understanding of the semantics of object orientation.

5   Knowledge and Competencies

Apparently, the two target groups of our concept map survey have extremely different educational backgrounds. While G8-students had 2-6 years of systematic CS education, the G9 students did not have any CS courses or eventually only courses without any curriculum. This assumption justifies the idea to define potential cognitive dimensions of programming knowledge by forming the pairwise intersection of the corresponding graph communities in Table 2. Using the intersections - and therefore only a subset of the original concepts - leads to a set of potential cognitive dimensions that may or may not be exhaustive for the given domain. On the upside, what remains in our subset has strong empirical backing.

Additionally, we have applied some interpretations in order to sharpen the dimensions by omitting the following concepts: *edges* and *nodes* in dimension 1 as parts of the *graph* concept, *record* and *table* in dimension 2 as representation structures of *data* in *working memory*, *class* in dimension 4, because all programs in purely object-oriented languages like Java are written as class definitions . Finally, in dimension 3, *association* was added despite its absence in the communities. In summary, we propose the following potential cognitive dimensions of programming competency:

Table 3. The proposed difficulty levels of programming competencies

Level	Dimension 3	Dimension 4
1	*Objects of only one class or of several unrelated classes*	*Single statements or sequences of statements*
2	*Objects of classes associated by 1:1 and 1:n associations*	*Single or sequentially arranged control structures*
3	*Objects of classes associated by m:n associations*	*Nested control structures*
Questionable	*Super- or subclasses, parametrized classes, recursive class structures (e.g. compositum pattern)*	*Methods, functions and procedures (MFP) without and with parameters, recursive MFPs, type parametrized MFPs*

1. Data structure (*graph, tree, array*)
2. Machine (*data, working memory, processor*)
3. Objects static structure (*object, attribute, association*)
4. Algorithmic structure (*loops, conditional statement*)
5. Representation structure (*programming language, syntax, semantics*)
6. Execution structure: (*statement, program, automaton*)

Regarding the potential difficulty levels on these dimensions, we decided to restrict our deliberations on the dimensions 3 and 4 for the moment, which seem to be particularly relevant for object oriented programming in practice. Suggested by the levels of the SOLO taxonomy (see above), we propose three levels for each of these dimensions (see Table 3). Of course, the hierarchy of these levels has still to be validated empirically. Additionally, for some other programming concepts the corresponding level seems questionable. Once the hierarchy of the levels 1-3 is proven, their difficulty has to be examined by suitable test items. Thus, we have collected these concepts them in a category "questionable". A first survey on dimension 4 resulted in very promising outcomes, which will be described soon in another paper.

6   Conclusion and Future Work

A look on our project outline shows that the definition and measurement of programming competencies is a very complex and challenging task, which requires many different research activities.

One of the first steps is the exploration of the cognitive structure of the considered competencies. Apparently, the evaluation of our concept mapping data suggests 6 different dimensions of this cognitive structure. Yet, it has still to be investigated by empirical means if these dimensions are representing separated psychometric constructs indeed. For this purpose, as a next step, suitable sets of items have to be constructed and validated (e.g. by expert interviews). Furthermore, these item sets have to be pretested, refined and finally applied in tests among some hundreds of test persons. By checking the homogeneity by means of the *Item Response Theory*, it will hopefully turn out how programming knowledge of novices is really structured. We have already undertaken the first step in this direction by designing, testing and evaluating a set of items for dimension 4.

Additionally, these quantitative investigations have to be supported by qualitative work. For this purpose, we are collecting learning diaries, reports about particular learning problems and many interviews from different target groups. Furthermore, we are investigating more than thousand assignments from K-12 schools and universities [27]. We have already designed and pretested a set of items for our proposed dimension 4 (see below) with very promising results, which will be published soon.

At the end, we hope to be able to assemble the results of all these activities to competency structure and level models that could form the starting point for the construction of standardized tests for programming abilities.

References

1. Bartholomew, D.J., Steele, F., Moustaki, I., Galbraith, J.I.: Analysis of Multivariate Social Science Data, 2nd edn. Chapman & Hall/CRC and CRC Press, Boca Raton (2008)
2. Bennedsen, J., Schulte, C.: A competence model for object-interaction in introductory programming. In: 18th Workshop of the Psychology of Programming Interest Group, vol. 18, pp. 215–229 (2006)
3. Bennedsen, J., Schulte, C.: Object interaction competence model v. 2.0. In: Learning and Teaching in Computing and Engineering, pp. 9–16. IEEE Press, Los Alamitos (2013)
4. Biggs, J.B., Collis, K.F.: Evaluating the quality of learning: The SOLO taxonomy; structure of the observed learning outcome. Educational Psychology Series. Acad. Pr., New York (1982)
5. Clauset, A., Newman, M.E.J., Moore, C.: Finding community structure in very large networks. Physical Review E 70(6), 066111 (2004)
6. Dearholt, D.W., Schvaneveldt, R.W.: Properties of pathfinder netowrks. In: Schvaneveldt, R.W. (ed.) Pathfinder Associative Networks, pp. 1–30. Ablex Pub. Corp., Norwood (1990)
7. Derbentseva, N., Safayeni, F., Cañas, A.J.: Concept maps: Experiments on dynamic thinking. Journal of Research in Science Teaching 44(3), 448–465 (2007)
8. Gouli, E.: Concept Mapping in Didactics of Informatics. Assessment as a Tool for Learning in Web-based and Adaptive Educational Environments. Ph.D. thesis, National and Kapodistrian University of Athens, Athen (2007)
9. Han, J., Kamber, M.: Data Mining: Concepts and Techniques. The Morgan Kaufmann series in data management systems, 2nd edn. Elsevier/Morgan Kaufmann, Amsterdam (2010)
10. Hawkins, W., Hedberg, J.G.: Evaluating logo: Use of the solo taxonomy. Australian Journal of Educational Technology 2(2), 103–109 (1986)
11. Hubwieser, P.: Computer science education in secondary schools – the introduction of a new compulsory subject. Trans. Comput. Educ. 12(4), 16:1–16:41 (2012)

12. Keppens, J., Hay, D.B.: Concept map assessment for teaching computer programming. Computer Science Education 18(1), 31–42 (2008)
13. Klieme, E., Avenarius, H., Blum, W., Döbrich, P., Gruber, H., Prenzel, M., Reiss, K., Riquarts, K., Rost, J., Tenorth, H.E., Vollmer, H.J.: The Development of National Educational Standards: An Expertise. Bundesministerium für Bildungund & Forschung, Berlin (2004)
14. Klieme, E., Hartig, J., Rauch, D.: The concept of competence in educational contexts. In: Hartig, J., Klieme, E., Leutner, D. (eds.) Assessment of Competencies in Educational Contexts, pp. 3–22. Hogrefe & Huber Publishers, Toronto (2008)
15. Magenheim, J., Nelles, W., Rhode, T., Schaper, N.: Towards a methodical approach for an empirically proofed competency model. In: Hromkovič, J., Královič, R., Vahrenhold, J. (eds.) ISSEP 2010. LNCS, vol. 5941, pp. 124–135. Springer, Heidelberg (2010)
16. Misra, S., Akman, I.: Measuring complexity of object oriented programs. In: Gervasi, O., Murgante, B., Laganà, A., Taniar, D., Mun, Y., Gavrilova, M.L. (eds.) ICCSA 2008, Part II. LNCS, vol. 5073, pp. 652–667. Springer, Heidelberg (2008)
17. Mühling, A.M.: Investigating Knowledge Structures in Computer Science Education. Ph.D. thesis, Technische Universität München, München (2014)
18. Neugebauer, J., Hubwieser, P., Magenheim, J., Ohrndorf, L., Schaper, N., Schubert, S.: Measuring student competences in german upper secondary computer science education. In: Gülbahar, Y., Karataş, E. (eds.) ISSEP 2014. LNCS, vol. 8730, pp. 100–111. Springer, Heidelberg (2014)
19. Novak, J.D.: Meaningful learning: the essential factor for conceptual change in limited or inappropriate propositional hierarchies leading to empowerment of learners. Science Education 86(4), 548–571 (2002)
20. Novak, J.D.: Learning, Creating, and Using Knowledge: Concept Maps as Facilitative Tools in Schools and Corporations, 2nd edn. Routledge, London (2010)
21. Novak, J.D., Cañas, A.J.: The theory underlying concept maps and how to construct and use them: Technical report IHMC cmaptools 2006–01 rev. 01–2008 (2008)
22. Novak, J.D., Cañas, A.J.: The universality and ubiquitousness of concept maps. In: Sánchez, J., Cañas, A.J., Novak, J.D. (eds.) Concept Maps: Making Learning Meaningful, vol. 1, pp. 1–13. Universidad de Chile, Chile (2010)
23. Novak, J.D., Musonda, D.: A twelve-year longitudinal study of science concept learning. American Educational Research Journal 28(1), 117–153 (1991)
24. OECD (ed.): PISA 2012 Results in Focus: What 15-year-olds know and what they can do with what they know. OECD Publishing, Paris (2013)
25. Ormerod, T.: Human cognition and programming. In: Hoc, J.M., Green, T., Samurcay, R., Gilmore, D.J. (eds.) Psychology of Prgramming. Computers and People, pp. 63–82. Academic Press, London (1990)
26. Rosas, S.R., Kane, M.: Quality and rigor of the concept mapping methodology: A pooled study analysis. Evaluation and Program Planning 35(2), 236–245 (2012)
27. Ruf, A., Berges, M., Hubwieser, P.: Types of assignments for novice programmers. In: Proceedings of the 8th Workshop in Primary and Secondary Computing Education, WiPSE 2013, pp. 43–44. ACM, New York (2013)
28. Ruiz-Primo, M.A., Shavelson, R.J.: Problems and issues in the use of concept maps in science assessment. Journal of Research in Science Teaching 33(6), 569–600 (1996)
29. Sanders, K., Boustedt, J., Eckerdal, A., McCartney, R., Moström, J.E., Thomas, L., Zander, C.: Student understanding of object-oriented programming as expressed in concept maps. ACM Inroads 40(1), 332–336 (2008)

30. Schvaneveldt, R.W., Durso, F.T., Dearholt, D.W.: Network structures in proximity data. The Psychology of Learning and Motivation 24, 249–284 (1989)
31. Seidel, T., Prenzel, M.: Assessment in large-scale studies. In: Hartig, J., Klieme, E., Leutner, D. (eds.) Assessment of Competencies in Educational Contexts, pp. 279–304. Hogrefe & Huber Publishers, Toronto (2008)
32. Shao, J., Wang, Y.: A new measure of software complexity based on cognitive weights. Canadian Journal of Electrical and Computer Engineering 28(2), 69–74 (2003)
33. Sousa, D.A.: How the Brain Learns: A Multimedia Kit for Professional Development, 3rd edn. Corwin Press, Thousand Oaks and Calif (2009)
34. Squire, L.R.: Memory and Brain. Oxford University Press, New York (1987)
35. Trumpower, D.L., Sharara, H., Goldsmith, T.E.: Specificity of structural assessment of knowledge. The Journal of Technology, Learning and Assessment 8(5) (2010)
36. Weinert, F.E.: Concept of competence: a conceptual clarification. In: Rychen, D.S., Salganik, L. (eds.) Defining and Selecting Key Competencies, pp. 45–65. Hogrefe and Huber, Seattle (2001)

Defining Proficiency Levels of High School Students in Computer Science by an Empirical Task Analysis

Results of the MoKoM Project

Jonas Neugebauer[1], Johannes Magenheim[1],
Laura Ohrndorf[2], Niclas Schaper[1], and Sigrid Schubert[2]

[1] University of Paderborn, D-33102 Paderborn, Germany
http://ddi.uni-paderborn.de
[2] University of Siegen, D-57068 Siegen, Germany
http://www.die.informatik.uni-siegen.de

Abstract. In the last few years an interdisciplinary team of researchers in the fields of organizational psychology and didactics of informatics have worked together to develop an empirical sound competence structure model, a measurement instrument and a competence level model. This is considered a relevant step for the reliable assessment of competences and the development of competence based curricula to foster the recent outcome orientation of the German educational system.

In this paper we publish the last component of our efforts: a model of proficiency levels, derived from the results of a competence assessment with over 500 German students. We describe different approaches to define proficiency levels and the process we used to derive them from our data. In the end, a detailed overview of the four proficiency levels is given and supplemented with exemplary tasks students should be able to solve on each.

Keywords. Competence Modeling, Proficiency Levels, Competence Level Model, Secondary Education.

1 Motivation

In 2013 we published a domain-specific competence structure model called "Competence model for informatics modeling and system comprehension" [1], which will be used to define educational standards and thereby contribute to the development of curricula and the measurement of competences and learning outcomes in diverse educational settings. Based on the competence model, a measurement instrument was created to assess competences of computer science students. This instrument was applied in an assessment of over 800 students in German upper secondary education (ages 16 to 19) to gather data for the evaluation of both the competence model and the instrument itself. The data analysis was done using the Multidimensional Item Response Theory (MIRT), a probabilistic test model. Based on these results, we concluded our research by developing a model of different proficiency levels, which is

© Springer International Publishing Switzerland 2015
A. Brodnik and J. Vahrenhold (Eds.): ISSEP 2015, LNCS 9378, pp. 45–56, 2015.
DOI: 10.1007/978-3-319-25396-1_5

related to the competence structure model and further on will allow us to examine the learning process of students in computer science education according to a competence development model.

This article will present the results of the last research step of the MoKoM project: the development of the competence level model, based on the analysis of difficulty determining characteristics of the test tasks. First, we will shortly summarize the previous results of the research outlined above. Then, we will give an overview of the theory behind proficiency levels, the ways they can be defined and the methods we used in the end. The article will conclude with the description the of four proficiency levels we derived from the data and supplement them with examples of tasks, which are distinctive for each level.

2   MoKoM Background and Prior Results

The MoKoM project – funded by the German Research Foundation (DFG) – is an interdisciplinary research project of psychologists and CSE-researchers to develop a competence model for the two domains of 'System Comprehension' and 'System Development'. Competence models are a necessary foundation for the creation of educational standards for individual subjects, a goal deemed necessary by the orientation of the German educational system towards competence based approaches in the recent years [2]. Developing these models relies on empirical methods to assess the required competences for important domains of a subject [1, 3].

The steps of the MoKoM project and the results of each phase are summarized in this section: First we developed a competence model, then we created test items and compiled them into a measurement instrument. The instrument was applied in an assessment of computer science students in German higher secondary education and the resulting responses were analyzed. Furthermore, an expert rating was conducted, to separately assess criteria for the objectively expected difficulty of each test item.

2.1   The MoKoM Competence Model

The development of the competence structure model had two phases. First, a theory based competence framework was derived from prominent curricula and syllabi, both national and international. This led to a framework encompassing four dimensions: Basic Competences, Informatics Views, Complexity and Non-Cognitive Competences (see [2, 4, 5]).

The framework was refined by conducting 30 expert interviews and analyzing the results by means of the Qualitative Content Analysis (see [4, 6]). This phase resulted in a empirically refined competence model with five dimension: K1 System Application, K2 System Comprehension, K3 System Development, K4 Dealing with Complexity and K5 Non-Cognitive Competences [1]. Each dimension consists of competence components, which in turn consist of individual competence facets. In the next step, test items for most of these facets were developed. The results were thoroughly discussed in several publications.

2.2   Test Instrument and Assessment Results

The test instrument contains 74 unrelated items to cover the MoKoM competence structure model [7]. Due to the large amount of items, the test instrument was not applicable in a classroom setting with timeslots of 90 minutes. In order to adapt the instrument to these demands, the items were divided into six booklets. The application of such a design is possible due to the MIRT, which allows the calculated estimation of student abilities in combination with the overall item difficulty [8]. Any subject has a certain probability to answer any item right or wrong. The difficulty of the item and the ability level of the subject determine this probability. MIRT models (as an extension of IRT models) assume, that multiple latent variables (one per dimension) cause the responses to a test [10]. IRT models allow the use of a matrix design with different booklets that only represent a part (about three-fourths) of the item pool of the competence test. The booklets were distributed to more than 800 computer science students in German upper secondary schools and the analysis of the 583 returned datasets allowed us to estimate the difficulty for each item. In 2014 we published the assessment results [9].

The main goal of this step was to examine the dimensional structure of the competence model and the reliability of the measurement instrument. Several different MIRT models, which assumed different dimensional structures of the test instrument, were used to analyze the empirical data. Afterwards the results were compared to assess the best fitting model. In our case the assumed complex competence structure containing the cognitive dimensions K1 to K4 with the additional non-cognitive dimension K5 proved to be the MIRT model with the best fit indices.

2.3   Model of Relevant Difficulty Criteria and Expert Rating

In the next step, we compiled a model of potentially relevant difficulty determining task characteristics from literature (e.g. from [11, 12] and [13]) and analyzed the items concerning informatics specific difficulty facets. On this basis thirteen features were identified and defined: addressed knowledge taxonomy level (KTL), cognitive process dimensions (CP), cognitive combination- and differentiation capacities (CCD), cognitive strain (CS), scope of tasks (necessary materials, reading effort and under-standing) (ST), inner- vs. outer computational task formulation, aspects of demands of computer science (IOC), number of components, level of connectedness (NC), stand-alone vs. distributed system (SDS), level of human-computer-interaction (HCI), (mathematical) combinatorial complexity (CC), level of the necessary understanding of systems of computer science (LUS), level of the necessary modelling competence of computer science (LMC). Each characteristic had a different number of facets (between 2 and 6) associated with it.

The measurement instrument was split into four parts of roughly equal size to keep the amount of ratings at an acceptable extent. Each of the four instrument parts was presented to two selected experts in the field of didactics of informatics, along with an explanation of each feature and its rating levels. The experts were asked to answer each item on their own, compare the solution with the given sample solution and then rate the item for each of the features. In addition, the experts had to give a subjective rating of the item difficulty on a scale from one to ten. The rating process resulted in a classification of the 74 items concerning each of the described difficulty determining characteristics.

To examine the relevance of the individual criteria on the item difficulty, the effect of the ratings on the difficulty estimates gained by means of the MIRT was determined through a linear regression analysis. This way the model of 13 criteria could be shrunk to a set of five with the most substantial influence on the item difficulty. These are KTL, CCD, CP, LUS and LMC. A detailed discussion of the difficulty criteria, processes used and results can be found in [14].

3   Derivation of Proficiency Levels

Based on the results outlined above, a competence level model was developed.

3.1   Definition of Proficiency Levels

Describing the abilities of a student within the probabilistic test theory is not as easy as it might seem. A students ability is characterized by the items he or she can solve, but this characteristic is only given as a probability. The item difficulty is described by a function of the student ability - the higher the students ability, the higher the probability to answer it correctly. In turn, for each student it is only known how likely it is he or she answers any given item correctly. For example in figure 1, the student s_3 is very likely to answer the item i_1 correct and the item i_2 wrong, whereas student s_4 has a high probability to solve both. Since students and items are arranged on a continuous scale, each student would have to be examined individually to evaluate his or her competences [15].

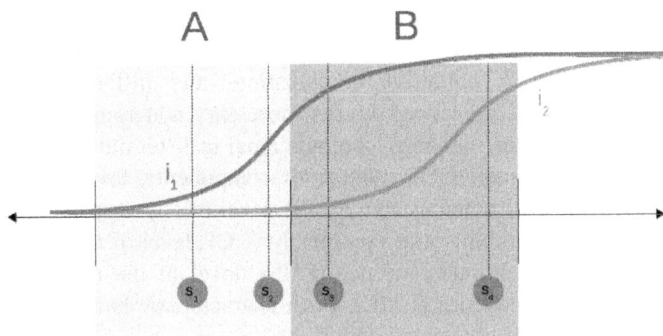

Fig. 1. Illustration of a continuous scale with two proficiency levels A and B, two item difficulty functions i_1 and i_2 and four students s_1 to s_4.

To be able to consistently assess and report competences of students, the definition and description of proficiency levels is required. These levels divide the continuous performance scale into discrete sections. This means that there are hard "cut-off" points, at which a student will be at one or the other level. These points are essentially arbitrary, though, since the probability of a student on level A (e.g. s_2 in figure 1) to solve an item (i_2) might be only slightly smaller than that of a student at the lower end of level B (s_3), while one placed at the upper end of level B (s_4) might have a much larger probability than both. Nevertheless, the definition of well-explained proficiency levels helps to understand the general distribution of competencies in the population and makes it easier to communicate the level of competence-gain over time (see [16, 17]).

3.2   Methods for the Description of Proficiency Levels

To derive proficiency levels from the results of a competence assessment, the most important aspect is the definition of thresholds or anchor points in the estimated ability that differentiate the individual proficiency levels. This method is called scale anchoring [18]. Defining these anchors can be done a-priori according to a predefined competence model, or post-hoc determined by the evaluated responses to a competence assessment. Large scale studies in the recent years like DESI [19], PISA [16] or TIMSS [20] have used different approaches to set thresholds and describe the proficiency levels.

There are multiple approaches to define anchor points that define the proficiency levels. The most pragmatic method is the definition based on rational arguments without an additional analysis of item characteristics. The thresholds can be set at predefined values, like the project NEAP for example scaled their data to be between 0 and 500 and used the anchor points 200, 250, 300 and 350. Other reasonable approaches (e.g. based on distribution percentiles) are valid, too. Then, the items with difficulty estimates close to one of the anchor points are chosen to describe each proficiency level. Usually items are selected if an adequate amount of examinees on a level was able to answer them correctly and an equally adequate amount of examinees on the level below the current one wasn't (for more details see [18]).

Another way to identify anchor points is concerned with the analysis of item characteristics to find striking items that suggest a change in the proficiency level, for example by requiring the use of more advanced skills for the first time. Table 1 shows three items (i_j) with their respective difficulty estimate (e_j), calculated from test responses by means of the MIRT, that illustrate this. Each item adds a previously missing criterion (C_k). Assuming each additional criterion increases the item complexity, these three items could also indicate an increased competence requirement.

Table 1. Illustration of change in item criteria.

Item	Estimate	C1	C2	C3
i_1	e_1	1	0	0
i_2	e_2	1	1	0
i_3	e_3	1	1	1

In the TIMSS study, the anchor points were selected based on the requirements to provide stable benchmark points for the international comparability over several years of assessments. The approach for selecting anchored items was similar to the direct anchoring method, but allowed for items to "almost anchor" by not fulfilling all anchor criteria completely. This was done to regard every item and not exclude those that would not fit the criteria. The description of the individual levels was done by a group of experts, who described the knowledge, understanding and skills each item required and formulated descriptions for each proficiency level from them [20].

In the PISA study, the cut-off points were determined by analyzing the items and following three principles: 1. The probability for a student to solve an item on the level should be at least 50 percent, 2. the width of each level should be roughly the same and 3. the probability to solve items of a higher / lower level should be higher / lower than 50 percent respectively. To achieve this, the PISA items were analyzed to describe how the proficiency requirements would increase. This was done by experts who linked aspects of the PISA framework to each item and generated an item map similar to table 1. By looking at the map they revealed patterns of criteria, which are related to item difficulty and which could be associated with locations on the continuous scale and thus describe proficiency levels [16].

3.3   Explorative Analysis of Results

To derive proficiency levels from the data gathered through the competence measurement instrument and the expert ratings (see sections 2.2 and 2.3) we chose an explorative approach., which included several steps.

Table 2. List of items ordered by difficulty estimates with a consecutive numbering, an internal item number, the internal item code, the estimate from the MIRT analysis, the difference in difficulty to the preceding item and five difficulty criteria as rated by experts (ranging from 0 to 6 depending on the criterion).

	Item	Label	Estimate	Diff	KTL	CCD	CP	LUS	LMC
22	...	...	...	...					
23	54	C6.A4.C1	500,9133	6,111	2	2	2	1	1
24	55	C6.A4.C2	500,9133	0,000	3	3	4	2	1
25	41	C5.A5	501,2106	0,297	3	2	5	2	1
26	46	C6.A2.B	513,424	12,213	1	1	2	1	0
27	6	C1.A6.B	519,4936	6,070	2	1	2	1	1
28	31	C4.A6	520,8621	1,369	3	2	4	2	1
29	...	...	...	...					

As a preparation, we combined the results into an overview of all items that were deemed of a high enough quality (based on the MIRT analysis of the responses) and the corresponding expert ratings. The list was ordered by the empirical difficulty estimate. The estimates were transformed according to the PISA standard, with a mean of 500 and standard deviation of 100. As a first approach, the expert ratings for each item were examined to identify which criteria would change at what point. For example whether a certain criterion wasn't present for all items up to a certain difficulty, but would appear on all items afterwards. This could indicate a change in proficiency levels. Unfortunately, the changes in the expert ratings were not clear enough to be interpreted as a change in proficiency. For example table 2 shows an excerpt from the item list. Looking at the five last columns, it is hard to see any striking changes in the ratings.

In a next step, the difficulty estimates were analyzed to reveal anchor points, where the estimates would jump by a significant amount. Starting with a difference of 20, we discovered several of such gaps. Unfortunately, most of these were located at the top or bottom of the list. The four easiest and the four hardest items displayed a great variance and inconsistencies concerning their difficulty. Defining every of these large differences as individual levels would not be suitable, because only one or two items would be available to describe the respective proficiency level.. It was decided to assign these extreme items commonly into the lowest and highest proficiency level respectively. By gradually decreasing the required difference in estimates, we identified five reasonable gaps with a minimum difference in difficulty of 12 points, which could serve as anchor points for five proficiency levels. Table 2 shows such an anchor point between row 25 and 26. Since the last level contained only five items with large differences in the difficulty estimate, it was decided to combine the last two levels into one to be able to describe it sufficiently. This decision was based on a close examination of all items in the highest two level in order to determine if the differences in required skills to solve the items justify this step.

In the next step, the four proficiency levels A to D had to be described according to the difficulty characteristics assigned to the items in each level. For this purpose, the scale anchoring method requires the selection of appropriate items near the anchor points, since those would be better suited to characterize the students' competences on this level (see section 3.2). To do so, we selected the six items adjacent to the anchor points and again examined and compared the difficulty characteristics for the three items on the lower level in contrast to those on the upper one. We hoped that changes in the characteristics would become clearer due to the specific anchor points,. Though this was possible for some of the levels, others were still hard to describe (e.g. the one in table 2). To deal with this, all items on two adjacent levels were included in the examination. By extracting commonalities between items on the same level and differences to the next lower one, each level could be described.

As a last step, we extended our analysis to the items itself. By looking at the content of each item, the proficiency level descriptions were refined and extended by sets of exemplary tasks a student on this level can be expected to solve.

4   Overview of Proficiency Levels

The proficiency levels were defined with increasing requirements from level A to level D. The full descriptions are shown in table 3.

Level A requires a very basic knowledge and most often involves everyday knowledge the students have gained through their prior experience with computer systems. One example task on this level is asking the student to name three components for input and two for output. That's undoubtedly a question one can answer without deeper computer science knowledge. However a change of perspective from a user, who simply uses these components, to a more technical view is required. Although this doesn't seem to be challenging at first, especially younger students have problems to describe the actual role and function of devices like a mouse (e.g. [21]). This task is a good example of how everyday knowledge is transferred to the school subject computer science and is enhanced by basic system comprehension.

Table 3. Descriptions of the four proficiency levels A to D.

Level	Description:
A	Learners are able to reproduce basic knowledge of informatics and use it to describe simple decisions concerning modeling, that are embedded in contexts close to everyday life. They are capable of comparing their everyday experiences to informatics systems
B	Learners are able to understand and explain simple decisions concerning modeling. They can illustrate functional interrelations of basic elements of informatics in bigger systems, as long as they belong to known contexts and are embedded in their everyday experience. Furthermore, they know basic terms, concepts of informatics and processes of object orientation and can explain them. Moreover, they can weigh up between these basic concepts and processes on the basis of their everyday experiences. They can make well-founded decisions and explain them with appropriate terminology. Additionally, a structural approach allows them to identify related fundamental ideas and concepts and to apply them in a given situation.
C	Learners are able to apply methods, concepts and processes of informatics in more complex contexts. They can weigh up different solution ideas and choose the most appropriate one by systematically analyzing the situation and understanding simple decisions of modeling. Moreover, they can develop solutions for problems by combining knowledge of different fields of informatics. At this level, they are also able to use advanced concepts and processes of object-oriented modeling and programming to analyze and modulate abstract system sequences.
D	Learners are able to apply their knowledge of fundamental ideas, concepts and abilities deriving from different fields of informatics to more generic problems and are able to explain the usage in abstract contexts. Moreover, they are able to study extensive analysis and design documents in order to understand new content. They are able to understand unknown terminology by interpreting it in the context of the problem. They can reflect on their (school) knowledge and are willing to expand their knowledge on their own.

Level B broadens the contexts and requires the students to be able to deal with more complex problem scenarios. This increasingly requires the identification and comparison of CS concepts and how they interact. Additionally, the students are able to use more precise terminology to describe systems. An example task on this level is an adequate response to the following question about object-oriented programming: "Describe the connection between a class and an object. Please use the possible class and objects 'teacher', 'Mr. Meyer' and 'Mr. Smith'."

Table 4. Exemplary tasks on each proficiency level.

Level	Example
A	They are able to create simple status diagrams based on descriptions of the functioning of systems close to everyday life. They can identify applications for different user roles of an IS and name criteria to evaluate them, for example criteria of software quality. They are able to identify and name parts of the user interface of the IS (hardware and software). They can name advantages and disadvantages of distributed systems.
B	They are able to name, understand and evaluate the quality of components and functions of web applications. They can implement and evaluate testing procedures for specifically predetermined program sections of selected parts of the system. They are able to name advantages and disadvantages of object-oriented programming languages and they can explain basic terms of object orientation (classes, objects, …) by giving concrete examples.
C	They are able to interpret and use definitions of requirements of systems to make and justify decisions (e.g. creating test cases, identifying components, …). On the basis of concrete examples, they can explain advanced terms of object orientation (inheritance, …) and use UML diagrams for object-oriented modeling (e.g. they transform applications into sequence diagrams). They are able to analyze systems with many components (e.g. booking system of a travel agency, library management), to differentiate between their functions and they can use their knowledge to solve problems (e.g. testing of the robustness of the system, creating new components).
D	They are able to work with APIs to use unknown classes within their implementation. They are able to identify classes based on written descriptions and model them as CRC cards. They are able to describe definitions and characteristics of object-oriented concepts on an abstract level. They are able to identify and classify the phases of the waterfall model and the fitting UML diagram types.

Similar to level A there still exists a connection to everyday life. However, it is transferred to a higher level of computer science knowledge by applying knowledge about ideas and concepts of informatics. Interestingly, this task is the easiest (regarding the test results) subtask of several questions concerning object-oriented programming. Especially two questions asking for the definition of "class" and "object" without the additional connection to examples from a known context had a much higher difficulty. This substantiates that it is a fundamental feature of this level, that tasks are embedded in a context students know from their everyday life.

A main progress from level B to C is the ability to move away from everyday examples and apply the knowledge to more general scenarios. An example for this is a task where a web-application for a travel agency including requirement definitions is described. The students have to complete three given drafts for test cases to test each of the requirements and three more to create unexpected cases that might produce an error in the application when entered. Besides the required expert knowledge, the test-taker now has to put him- or herself in the role of a professional software developer. However, there is still a link to everyday knowledge since the software serves a purpose students are familiar with (booking a holiday trip).

Level D is the highest proficiency level and therefore requires that students are able to apply their knowledge in diverse situations and contexts and connect this knowledge. One example is the modeling of CRC-cards based on the textual description of a problem situation. In this task, the students need to identify classes from the text and extract their functions as well as connections among each other. This requires factual knowledge about CRC cards and object oriented concepts on the one hand, and on the other hand the ability to apply this rather abstract knowledge in a real problem scenario and use structured approaches to extract relevant information from a written description. Table 4 gives an overview of exemplary tasks, which students on each level should be able to solve.

5   Conclusions

With this article and the upcoming publication of our book the last part of our research efforts towards an empirical based competence model is completed. We were able to substantiate the further development of competence-based curricula for (German) computer science education. Furthermore, the MoKoM measurement instrument and the proficiency levels allow the assessment of students' competences in large groups as well as the comparison of competence achievement over time. The description of the proficiency levels alongside the exemplary tasks should allow teachers to get a sense of the competences their students have and how they can design their lessons to facilitate competence development.

However, like any research project, there are a lot of ways to continue our research and follow up on our results:

- A measurement like ours can never be seen as a 'final' version. Though it may have a stable and usable state, it is advisable to do further evaluations and revisions of the test items, for a better fitting to the requirements. There are some items, that proved to be inadequate for our assessments based on statistical reference values (see [9]). Those items especially should be analyzed and reworked.

- Out competence model focuses on two domains: 'System Comprehension' and 'System Modeling'. Our experiences with the development process of the model could serve as a guideline to extend the model to further domains of computer science education.

- Our measurement instrument (and future revisions) may be used to assess competences of computer science students on a large scale and over time, to assess the current state of computer science education in Germany and identify areas to improve upon.

- To foster competence centered learning environments, our proficiency levels should be supplemented by specific learning scenarios, which aim to elevate students with regard to the higher levels of the scale. These scenarios focus on the processes required to process through the proficiency levels and form a competence development model that describes criteria how to assist the transition between levels.

- The existence of an empirical derived competence model and the results of other empirical based CSE-research projects should encourage the revision of existing standards for computer science education. We attempted a first step towards this with the comparison of the German Informatics Standards to the MoKoM competence model in [22], but further discussion and development is required to obtain satisfying results.

References

1. Linck, B., Ohrndorf, L., Schubert, S., Stechert, P., Magenheim, J., Nelles, W., Neugebauer, J., Schaper, N.: Competence model for informatics modelling and system comprehension. In: Proceedings of the 4th Global Engineering Education Conference, IEEE EDUCON 2013, Berlin, pp. 85–93 (2013)
2. Neumann, K., Fischer, H.E., Kauertz, A.: From Pisa To Educational Standards: The Impact Of Large-Scale Assessments On Science Education In Germany. International Journal of Science and Mathematics Education 8, 545–563 (2010)
3. Koeppen, K., Hartig, J., Klieme, E., Leutner, D.: Current Issues in Competence Modeling and Assessment. Zeitschrift für Psychologie 216, 61–73 (2008)
4. Magenheim, J., Nelles, W., Rhode, T., Schaper, N., Schubert, S., Stechert, P.: Competencies for informatics systems and modeling: results of qualitative content analysis of expert interviews. In: Proceedings of the 1st Global Engineering Education Conference, Educon 2010, pp. 513–521. IEEE Computer Society, Madrid (2010)
5. Nelles, W., Rhode, T., Stechert, P.: Entwicklung eines Kompetenzrahmenmodells – Informatisches Modellieren und Systemverständnis. Informatik-Spektrum 33, 45–53 (2009)
6. Lehner, L., Magenheim, J., Nelles, W., Rhode, T., Schaper, N., Schubert, S., Stechert, P.: Informatics Systems and Modelling – Case Studies of Expert Interviews. In: Reynolds, N., Turcsányi-Szabó, M. (eds.) KCKS 2010. IFIP AICT, vol. 324, pp. 222–233. Springer, Heidelberg (2010)

7. Rhode, T.: Entwicklung und Erprobung eines Instruments zur Messung informatischer Modellierungskompetenz im fachdidaktischen Kontext (Doctoral dissertation). University of Paderborn (2013)

8. Osteen, P.: An Introduction to Using Multidimensional Item Response Theory. Journal of the Society for Social Work and Research 1, 66–82 (2010)

9. Neugebauer, J., Hubwieser, P., Magenheim, J., Ohrndorf, L., Schaper, N., Schubert, S.: Measuring Student Competences in German Upper Secondary Computer Science Education. In: Gülbahar, Y., Karataş, E. (eds.) ISSEP 2014. LNCS, vol. 8730, pp. 100–111. Springer, Heidelberg (2014)

10. Hartig, J., Magenheim, J., Höhler, J., Nelles, W., Rhode, T., Schaper, N., Schubert, S., Stechert, P.: Multidimensional IRT models for the assessment of competencies. Studies in Educational Evaluation 35, 57–63 (2009)

11. Anderson, L.W., Krathwohl, D.R.: A Taxonomy for Learning, Teaching, and Assessing: A Revision of Bloom's Taxonomy of Educational Objectives. Addison Wesley Longman, New York (2001)

12. Seehorn, D., Carey, S., Fuschetto, B., Lee, I., Moix, D., O'Grady-Cunniff, D., Boucher Owens, B., Stephenson, C., Verno, A.: CSTA K–12 Computer Science Standards. Computer Science Teachers Association Association for Computing Machinery (2011)

13. Schaper, N., Ulbricht, T., Hochholdinger, S.: Zusammenhang von Anforderungsmerkmalen und Schwierigkeitsparametern der MT21-Items. In: Blömeke, S., Kaiser, G., and Lehmann, R. (eds.) Professionelle Kompetenz angehender Lehrerinnen und Lehrer: Wissen, Überzeugungen und Lerngelegenheiten deutscher Mathematikstudierender und-referendare. Erste Ergebnisse zur Wirksamkeit der Lehrerausbildung, pp. 453–480. Waxmann Verlag (2008)

14. Magenheim, J., Nelles, W., Neugebauer, J., Ohrndorf, L., Schaper, N., Schubert, S.: Expert rating of competence levels in Upper Secondary Computer Science Education. In: Brinda, T., Reynolds, N., and Romeike, R. (eds.) KEYCIT 2014 – Key Competencies in Informatics and ICT, pp. 1–12 (2014).

15. Watermann, R., Klieme, E.: Reporting Results of Large-Scale Assessment in Psychologically and Educationally Meaningful Terms. European Journal of Psychological Assessment 18, 190–203 (2002)

16. Organisation for Economic Co-operation and development: PISA 2003 Technical Report. OECD (2005)

17. Hartig, J., Frey, A., Nold, G., Klieme, E.: An Application of Explanatory Item Response Modeling for Model-Based Proficiency Scaling

18. Beaton, A.E., Allen, N.L.: Interpreting Scales Through Scale Anchoring. Journal of Educational and Behavioral Statistics 17, 191–204 (1992)

19. DESI-Konsortium [Hrsg]: Unterricht und Kompetenzerwerb in Deutsch und Englisch Ergebnisse der DESI-Studie Weinheim ua: Beltz, pp. 34–54 (2008)

20. Martin, M.O., Mullis, I.V.S.: Overview of TIMSS 2003. In: Martin, M.O., Mullis, I.V.S., Chrostowski, S.J. (eds.) TIMSS 2003 Technical Report, pp. 3–20. Boston College, Chestnut Hill (2004)

21. Hammond, M., Rogers, P.: An investigation of children's conceptualisation of computers and how they work. Education and Information Technologies 12, 3–15 (2006)

22. Magenheim, J., Neugebauer, J., Stechert, P., Ohrndorf, L., Linck, B., Schubert, S., Nelles, W., Schaper, N.: Competence Measurement and Informatics Standards in Secondary Education. In: Diethelm, I., Mittermeir, R.T. (eds.) ISSEP 2013. LNCS, vol. 7780, pp. 159–170. Springer, Heidelberg (2013)

Classification of Programming Tasks According to Required Skills and Knowledge Representation

Alexander Ruf, Marc Berges, and Peter Hubwieser

Technische Universität München, TUM School of Education
Arcisstr. 21, 80333 München, Germany
{alexander.ruf,berges,peter.hubwieser}@tum.de

Abstract. Tasks represent a central part of computer science lessons, and aim to practice programming skills or to concrete abstract concepts for example. We have investigated, which types of tasks are given to novice programmers, typically. For that purpose, we have analyzed and generalized tasks from textbooks and exercise sheets. The result is a list of twelve task types classified according to required skills and knowledge representation. In addition, we found that the task types differ very much regarding their incidence. Finally, we tried to relate the three found forms of knowledge representation to concepts of cognitive psychology.

Keywords: task, task type, programming task, programming, novice programmer, classification, knowledge, knowledge representation, representation mode, cognitive process, skill, cognitive psychology, computer science education.

1 Introduction

Tasks play a key role in computer science education. Students perform tasks in the classroom given to them as examples or exercises, at home as homework or in tests to assess their performance. Teachers in turn create these tasks or select tasks from a collection, correct the solutions of their students and use the results to gain insight into how their courses are doing. Not surprisingly, tasks have always been the subject of research projects. This paper deals with the classification of tasks. Tasks can be classified by very different properties, such as the difficulty, the purpose, the type or form (e.g. open or multiple choice questions), or the context, to mention only some features. We try to classify tasks according to their type. Moreover, we focus on programming tasks, provided to novice programmers. The type of a task is defined by the skills required to solve the task and forms of knowledge representation in the given problem and in the expected solution, respectively. We will describe our method in detail in section 3. The task types we could identify are then presented in section 4. In addition, we wanted to know, how much the task types that we found in the selected sources differ in their incidence. These results are presented in section 5. And finally, in section 6 we suggest a possible relation between the forms of knowledge representation that we found in our tasks and concepts of cognitive psychology.

© Springer International Publishing Switzerland 2015
A. Brodnik and J. Vahrenhold (Eds.): ISSEP 2015, LNCS 9378, pp. 57–68, 2015.
DOI: 10.1007/978-3-319-25396-1_6

2   Related Work

In a study on gender differences in preferring a specific type of assignment, Wilson stated three categories of assignments (see [23]). The categories are oriented on their field of application. Three types are declared that are typical for CS1 textbooks: "Real-world" problems, games and mathematical problems. The students of the investigation are asked to give their preferences on the kind of assignment. Layman et al. describe in [17] the nature of programming assignments. They categorize assignments of introductory CS1 and software engineering courses of selected institutions into five categories. Besides a category for assignments around "Karel the Robot" (see [18]), the programs are ordered into games, programs with little practical context, programs with practical context or programs with social relevance. Their findings showed that most of the investigated assignments have no practical context. Another investigation of programming projects gets along with the same purpose. Hansen [13] asked students on their assessment concerning the niftiness of assignments. For analysis the assignments itself are categorized by topics of the underlying course materials. This categorization is quite ambiguous and leads to several problems mentioned by the author. The description and classification of programming tasks was even done by Tharp in 1981 when there was a shift from learning syntax to teaching the concepts behind CS. In [22] he defines categories by programming goals and gives examples for each category.

Nielsen et al. built a taxonomy of questions in a two dimensional model. One dimension is built by a list of question types, the other dimension contains different information on the task like the top level of Blooms taxonomy or form of the response.

In his doctoral thesis (see [7]) Brinda has examined task classes for object-oriented modeling very comprehensively and completely and embedded them in a larger context. In addition to the formation of classes, he established a connection to taxonomies and investigated the context of tasks for object-oriented modeling.

Bower describes in [4] a taxonomy of task types in computing, deductively derived from curricula and literature and verified by experts. Hazzan et al. in [14] and Ragonis in [19], where she published the question types from [14] for the olympiads in informatics, present deductively derived questions as a resource for teachers.

In the context of the BRACElet project Sheard et al. have developed a classification scheme for examination questions; they presented it in [21]. In this scheme, a distinction is made between six categories, e.g. "topics covered" or "skills required to answer the question" (see also subsection 3.3).

3   Methodology

3.1   Sources for the Tasks

Before analyzing any tasks we have to get a list of sources. As in [20] already described, the tasks analyzed should cover nearly all aspects concerning the

population that is addressed by the tasks (students or self-learning readers).
Besides that the kind of authors (e.g. teachers at school and university or authors
of textbooks) should be covered as well. Therefore we investigated three groups
of texts. Neither the groups nor the elements in the group are exhaustive. But
for reasons of coverage they are sufficient.

The first group contains secondary school textbooks in Bavaria. The compulsory subject is described in detail in [15]. All students learn the basic concepts
of object-orientation and programming in their first years at school (grade six
and seven). If they continue with the combination nature and technology they
have an introductory course into object-oriented programming in grade ten.
For this reason we include the books of these grades. For each of these grades
there are two different books that are used in school, so we have a total of four
books: [12], [5], [16], and [6].

From the university context, we selected the introductory course of our university for practical reasons. Besides the lecture, every student (major and non-
major) must participate in a practical course on object-oriented programming.
The worksheets with the tasks are included in our analysis. Although the main
part of the freshmen at our university comes from Bavaria, the practical course
does not ask for any programming experiences.

In a paper on visualizing conceptual knowledge in texts (see [3], [2]), we did an
investigation of textbooks in introductory programming courses in national and
international universities. We counted the textbooks recommended in the courses
of the universities in the list and selected those which were not singleton. For
the recent investigation we chose the two books [10] and [11] from the resulting
list with the most concepts related to object-orientation (see [3]).

3.2   Selection of the Tasks

The chosen books contain a huge amount of tasks. Due to the fact that we only
want to analyze tasks for novice programmers, we concentrate on programming
tasks. Besides the different topics the tasks are related to, the length and scope
of the tasks is quite spreading. We include all tasks in our analysis that contain
any programming code either in the given problem or in the expected solution.
The extent of the programming code does not matter, starting with just a line
of code to the full program everything is included.

The sources referred to in subsection 3.1 use Java as their programming language – with exception of the two textbooks of the grades 6/7, they work with
a German version of "Karel the robot" (see [18]), a programming environment
with a "mini language" (see [9]), designed specifically for young novice programmers. In accordance with the Bavarian curriculum for the 7th grade (see [15]) an
object-oriented programming approach is also adopted there. So, all of the tasks
in this paper have been developed for novice programmers who learn object-
oriented programming.

Often, a task in the sources consists of several parts. Since the partial tasks
very often differ in type we have treated and examined each subtask as an own
task in these cases.

We have restricted ourselves to tasks for novice programmers, but two text-books go far beyond, we had to restrict the content meaningfully. Since the Bavarian curriculum for grade 10 reaches to the topics inheritance and polymorphism [15], we have included tasks from the two textbooks in our investigation only up to these topics, too.

3.3   Classification of the Tasks

In a first study (see [20]), we captured what is given in the respective task and what the student has to do to solve the task. Then we stripped both criteria "given" and "to-do" from the context and formulated them in a generic way. Two tasks were classified in the same category if they have basically the same given and if the same is to do.

More complex tasks, which involved more than one "to-do" were divided into corresponding parts and associated with multiple categories, i.e. an "atomic" task was made from each "to-do", which was then used for further investigation.

For each of these "atomic" tasks it can be distinguished between knowledge and cognitive process dimension comparable with learning objectives as suggested by Anderson and Krathwohl in their revised Bloom's taxonomy [1]. Knowledge elements in tasks occur both in the given problem and in the expected solution whereas the cognitive process hides in the description of what is to do (see Fig. 1).

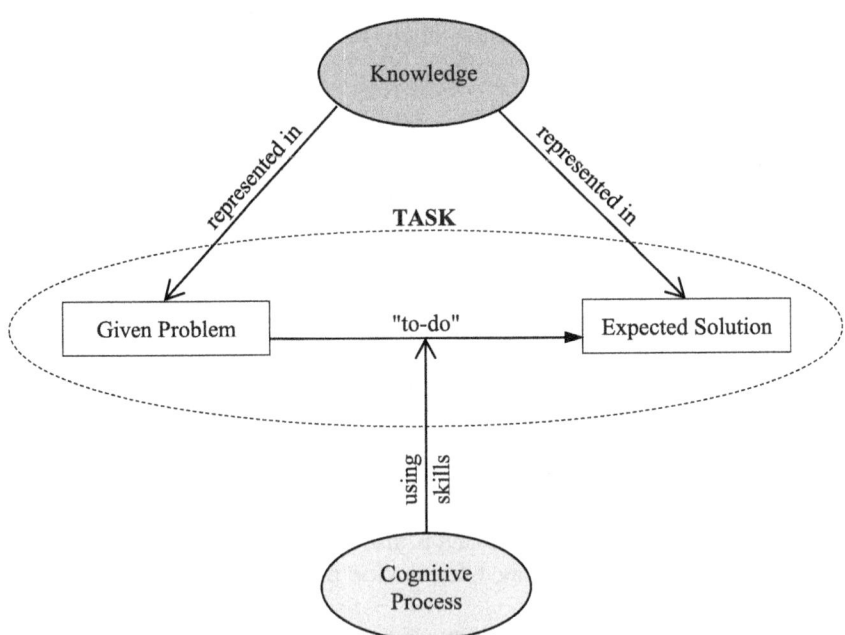

Fig. 1. The roles of knowledge and cognitive process while solving a task

Of course number and type of knowledge elements differ very from task to task, as well different cognitive processes run depending on the task. Seen in this way it seems that these dimensions are inappropriate for our intention to classify tasks. However, the tasks differ little in the form, in which the knowledge is represented in the given problem and in the expected solution, respectively. Also the already extracted "to-do"s vary little in the underlying skills, so in a sense they can be regarded as a generic term or a "headline" for the various cognitive processes, similar to the corresponding category in the classification scheme for examination questions in [21].

So we read out the skills from the "to-do"s and deduced the forms of knowledge representation in the given problem and in the expected solution, respectively, from the "given" and the "to-do"s. After that we classified the tasks according to the skills required to solve the task and the forms of knowledge representation.

The result is presented in the following section.

4   The Resulting Types

Overall 967 tasks, fulfilling the criteria of subsection 3.2, were found in the seven sources referred to in subsection 3.1. From these 967 tasks 1098 "atomic" tasks as described in subsection 3.3 resulted. Within these 1098 tasks we could identify 11 different types, however, one type splits in two different subtypes. What exactly is hidden behind each type is explained below. The title therein specifies the required skill and under the item "forms of knowledge representation" the possible representation form(s) in the given problems are to the left of the arrow and the possible representation form(s) in the expected solutions to the right.

4.1   Type 1a. "write code"

Forms of knowledge representation: **Text** → **Code**

Example: A worldwide operating business is involved in several companies and subsidiaries. Each of these companies is divided in departments, which again consists of several employees. Implement the involved classes. (From [16], translated and adapted by the authors.)

Notes: The given problem is usually (more or less) openly formulated, how much the problem is described in detail, however, varies considerably. The context varies widely, too.

4.2   Type 1b. "write code"

Forms of knowledge representation: **Diagram** → **Code**

Example: Implement the given state diagram of a shaver. (From [16], translated and adapted by the authors.)

Notes: The given diagram types can be very different, for example class diagram, object diagram, state diagram, or sequence diagram.

4.3   Type 2. "write code using the given code"

Forms of knowledge representation: **Text and Code → Code**

Example: Write a new method `buildChessboard()` using the given method
`placeChessrow()`. (From [12], translated and adapted by the authors.)

Notes: In addition to a problem, comparable to that of type 1a, requirements
regarding the solution are made. These requirements specify e.g. the
use of given methods or classes.

4.4   Type 3. "adjust/extend/complete the given code"

Forms of knowledge representation: **Text and Code → Code**

Example: a) Following the HelloDate.java example, create a "hello, world" program, that simply displays that statement.
b) Turn the DataOnly code fragment into a program that compiles
and run. (Both examples from [11].)

Notes: In some cases this type could be regarded as special case of type
2, because the transition from "write code using the given code" to
"extend the given code" is smooth, for example.

4.5   Type 4. "optimize the given code"

Forms of knowledge representation: **Text and Code → Code**

Example: Consider the points in the given solution at which it could be useful
to have random elements. Transfer your consideration to the program.
(From [6], translated and adapted by the authors.)

Notes: Note the difference to type 3. Usually there is only the solution to a
similar problem or to a part of the problem given, whereas here the
solution is matching to the given problem and has just to be optimized.
Further note, that not only the improvement of the code is meant but
the improvement of the solution with regard to the given problem.

4.6   Type 5. "debug the given code"

Forms of knowledge representation: **Code (and Text) → Code (and Text)**

Example: Identify and correct the errors in the following piece of code.

```
while ( y < 10 )
    System.out.println( a );
    --a;
    }
```

(From [10].)

Notes: Often in this task type it is not given if the given code is correct or not. In this case the student first has to decide if the given code is correct or not and then has to correct it or give reasons for it. It is also possible that multiple, often similar solutions are given, that all should be checked for correctness. Also pure syntax corrections without any reference to a "real" problem, as in the example above are possible.

4.7   Type 6. "set the right preconditions to the given code"

Forms of knowledge representation: **Text and Code → Text**

Example: Design a labyrinth (with at least one exit), from which Karel never finds out using the given method `searchExit()`. (From [12], translated and adapted by the authors.)

Notes: As the example shows, says "the right preconditions" not always that the program must terminate.

4.8   Type 7. "test the given code"

Forms of knowledge representation: **Code → Code**

Example: Test your implementation with a small example program. (From the students' worksheets, translated and adapted by the authors.)

Notes: Often, it is more precisely specified in these tasks, how the test should be performed. These include, for example, information to the program inputs or the objects that are to be created interactively. The program code to be tested comes sometimes from previous programming tasks.

4.9   Type 8. "transform the given code"

Forms of knowledge representation: **Code → Code**

Example: Convert the following for loop to a while loop:

```
for (int x = 50; x > 0; x--)
    {
        System.out.println(" x = " + x);
    }
```

(From [10].)

Notes: "Transform the given code" even contains the transformation into a different programming language, including pseudo code.

4.10   Type 9. "trace/explain the given code"

Forms of knowledge representation: **Code → Text**

Example: The following program shows an implementation of an algorithm for the greatest common divisor of two numbers. Test the implementation by writing the execution of the method calls gcd(35, 20) and gcd(35, -7) in a table.

```
public int gcd(int a, int b) {
    while (a != b) {
        if (a >= b) {
            a = a ? b;
        }
        else {
            b = b ? a;
        }
    } return a;
}
```

(From [16], translated and adapted by the authors.)

Notes: In this task type the student has to trace code execution mentally. Often program inputs are given, too. The product of the mental code execution can be very different, for example:

- Describe the algorithm, that underlies the program code, in own words.
- Specify the program outputs / object states after the code execution.
- Create a tracing table.
- Specify the instanced objects during code execution.
- Document the given code.
- Specify how often a particular loop is executed.
- Explain the meaning of certain code elements.

4.11   Type 10. "specify a problem to the given code"

Forms of knowledge representation: **Code → Text**

Example: Consider to each of the given constructors a meaningful application. (From [6], translated and adapted by the authors.)

Notes: Certainly more specifications can be made, for example to the context. This task type is in some extend the inverse of type 1a.

4.12   Type 11. "draw a diagram to the given code"

Forms of knowledge representation: **Code → Diagram**

Example: Draw the control flow graph for the following MiniJava program:

```
int x, r, n;
r = 1; n = 1; x = read();
while ( n < x ) {
    if ( r % 1 == 0 )
        r = r * n;
    else
        r = r * (-n);
    n = n + 1;
    write( r );
}
```

(From the students' worksheets, translated and adapted by the authors.)

Notes: Of course the type of the diagram is specified closer in the respective task. Examples of possible diagrams are similar to type 1b. This task type is in some extend the inverse of type 1b.

5   Incidence of the Types

Considering Fig. 2, first of all it is striking, that the incidence of each type is very different.

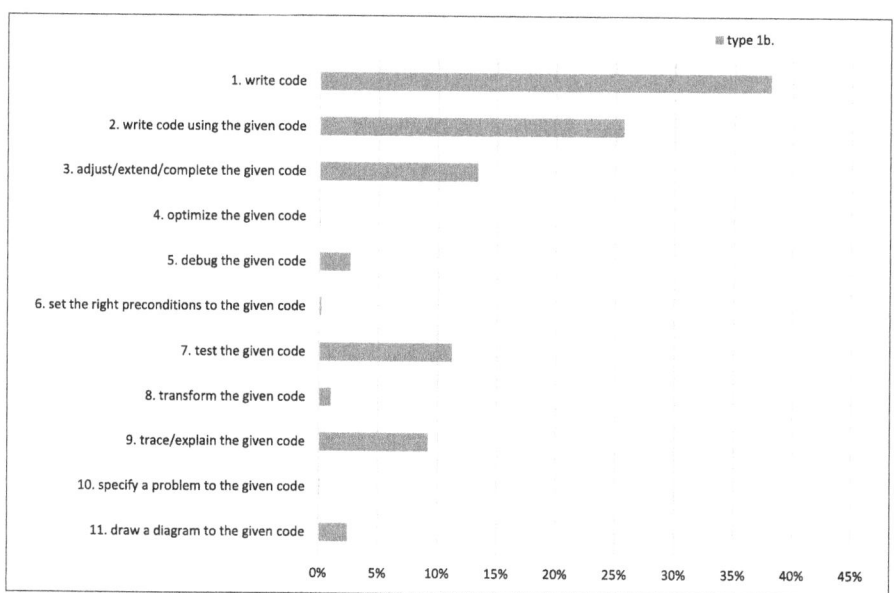

Fig. 2. Incidence of the types

On the five types 1, 2, 3, 7, and 9 distribute more than 90%, while the four types 4, 6, 8, and 10 constitute less than 2%. The single sources differ barely from this impression.

More in-depth statistical analysis of the data obtained would lead too far at this point. The objective of the analysis was actually only the identification of the types. Sometimes, the transition from one type to another is smooth, e.g. between the types 2 and 3. In this case the classification of a task can depend on the previously edited tasks of the student, so that the type is not objectively determinable.

Our results correspond very well with the results in [21]. "write code" predominates there in the category "skill required", too. That "trace code" is found more often there and "test code" less, this is probably because of the different sources, in [21] only examination questions were examined.

6   Forms of Knowledge Representation in the Tasks

In all of the examined tasks only three different forms of knowledge representation occur: text, code, and diagram.

Between these three types various combinations and transitions are possible as the description of the different types shows (see section 4). It is remarkable that these three types may be associated with the three modes of representations by J. Bruner [8], one of the founder of the cognitive psychology and a pioneer in the research field of knowledge representation.

Bruner distinguishes between enactive, iconic, and symbolic representation. The enactive representation is considered as action-based, the iconic representation as image-based, and the symbolic representation as language-based. Therefore the knowledge representation form "text" can be seen as corresponding to the symbolic representation, "diagram" as corresponding to the iconic representation, and "code" as corresponding to enactive – at least if the task is edited on a computer.

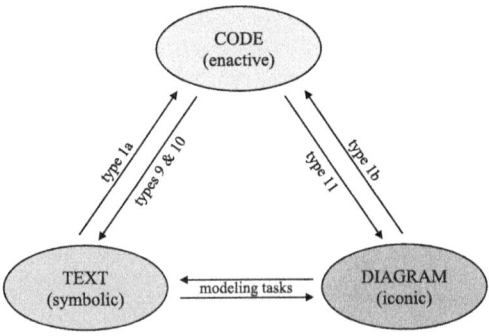

Fig. 3. E-I-S principle: The three modes of representation corresponding to the three forms of knowledge representation found in the tasks

According to Bruner, the transformation of knowledge from one form of representation to another is for cognitive development particularly "valuable" (intermodal transfer). Such transitions take place in the types 1a, 1b, 9, 10, and 11 (see Fig. 3).

For a holistic learning experience the transitions between "text" and "diagram" would have to occur in tasks, too. Since these tasks are purely modeling tasks they were not considered in this paper (see subsection 3.2).

7 Conclusion and Future Work

First of all, we hope, that our list of task types helps teachers to vary the tasks they use in their programming lessons. This list is fairly complete as the comparison with deductively derived types shows. For example, every type presented in [4] or [14] is transferable to our classification, provided that it refers to programming tasks.

Another point is that our investigation showed that the sources we have selected vary little in their range of types in the offered tasks. We don't think, that the reason for this lies in our selection of the sources, rather the authors of the tasks don't exhaust the full diversity of the types. Maybe our list of task types can help authors of textbooks to vary task types consciously.

For the future it would be interesting to investigate if it is possible to classify the single task types into the revised Bloom's taxonomy. Of course, no type would be placed in only one cell of the scheme of Anderson and Krathwohl, but maybe there would be cells containing more task types than other.

In order to measure the programming competencies of students, test tasks of all possible types are required. Moreover, the classification of tasks is a prerequisite to be able to compare different test tasks. So, our list can help to develop such test tasks. But before programming competencies have to be identified. And there task types could help, too. You see, tasks and their classification are still challenging fields for research.

References

1. Anderson, L.W., Krathwohl, D.: A Taxonomy for Learning, Teaching, and Assessing: A Revision of Bloom's Taxonomy of Educational Objectives. Longman, New York (2001)
2. Berges, M., Hubwieser, P.: Towards an overview map of object-oriented programming and design. In: Proceedings of the 12th Koli Calling International Conference on Computing Education Research, Koli Calling 2012, pp. 135–136. ACM, New York (2012)
3. Berges, M., Hubwieser, P.: Concept specification maps: displaying content structures. In: Proceedings of the 18th ACM Conference on Innovation and Technology in Computer Science Education, ITiCSE 2013, pp. 291–296. ACM, New York (2013)
4. Bower, M.: A taxonomy of task types in computing. In: Proceedings of the 13th Annual Conference on Innovation and Technology in Computer Science Education, ITiCSE 2008, pp. 281–285. ACM, New York (2008)

5. Brichzin, P., Freiberger, U., Reinold, K., Wiedemann, A.: Ikarus, Natur und Technik, Schwerpunkt Informatik 6/7, 2nd edn. Oldenbourg, München (2005)
6. Brichzin, P., Freiberger, U., Reinold, K., Wiedemann, A.: Informatik II, Objektorientierte Modellierung. Oldenbourg, München, 1. Auflage (2008)
7. Brinda, T.: Didaktisches System für objektorientiertes Modellieren im Informatikunterricht der Sekundarstufe II (Dissertation). Fachbereich Elektrotechnik und Informatik, Universität Siegen (2004)
8. Bruner, J.S., Olver, R.R., Greenfield, P.M.: Studies in cognitive growth (1966)
9. Brusilovsky, P., Calabrese, E., Hvorecky, J., Kouchnirenko, A., Miller, P.: Minilanguages: a way to learn programming principles. Education and Information Technologies 2(1), 65–83 (1997)
10. Deitel, P.J., Deitel, H.M.: Java: How to program, 9th edn. Prentice Hall, New Jersey (2012)
11. Eckel, B.: Thinking in Java, 4th edn. Prentice Hall, New Jersey (2006)
12. Frey, E., Hubwieser, P., Winhard, F.: Informatik: Objekte, Strukturen, Algorithmen, Schülerbuch - Jahrgangsstufen 6 und 7. Klett, Stuttgart (2004)
13. Hansen, S.A.: Analyzing programming projects. In: Proceedings of the 40th ACM Technical Symposium on Computer Science Education, SIGCSE 2009, pp. 377–381. ACM, New York (2009)
14. Hazzan, O., Lapidot, T., Ragonis, N.: Guide to teaching computer science: An activity-based approach. Springer, New York (2011)
15. Hubwieser, P.: Computer science education in secondary schools - the introduction of a new compulsory subject. Trans. Comput. Educ. 12(4), 16:1–16:41 (2012)
16. Hubwieser, P., Spohrer, M., Steinert, M., Voß, S.: Algorithmen, objektorientierte Programmierung, Zustandsmodellierung, Schülerbuch - Jahrgangsstufe 10. Klett, Stuttgart (2008)
17. Layman, L., Williams, L., Slaten, K.: Note to self: make assignments meaningful. In: Proceedings of the 38th SIGCSE Technical Symposium on Computer Science Education, SIGCSE 2007, pp. 459–463. ACM, New York (2007)
18. Pattis, R.E., Roberts, J., Stehlik, M.: Karel the robot: A gentle introduction to the art of programming, 2nd edn. Wiley, New York (1995)
19. Ragonis, N.: Type of questions - The case of computer science. Olympiads in Informatics 6, 115–132 (2012)
20. Ruf, A., Berges, M., Hubwieser, P.: Types of assignments for novice programmers. In: Proceedings of the 8th Workshop in Primary and Secondary Computing Education, WiPSE 2013, pp. 43–44. ACM, New York (2013)
21. Sheard, J., Simon, C.A., Chinn, D., Laakso, M.-J., Clear, T., Raadt, M.: d., D'Souza, D., Harland, J., Lister, R., Philpott, A., Warburton, G.: Exploring programming assessment instruments: a classification scheme for examination questions. In: Proceedings of the Seventh International Workshop on Computing Education Research, ICER 2011, pp. 33–38. ACM, New York (2011)
22. Tharp, A.L.: Getting more oomph from programming exercises. In: Proceedings of the Twelfth SIGCSE Technical Symposium on Computer Science Education, SIGCSE 1981, pp. 91–95. ACM, New York (1981)
23. Wilson, B.C.: Gender differences in types of assignments preferred: implications for computer science instruction. Journal of Educational Computing Research 34(3), 245–255 (2006)

Online vs Face-To-Face Engagement of Computing Teachers for their Professional Development Needs

Sue Sentance[1] and Simon Humphreys[2]

[1] King's College London, UK
sue.sentance@kcl.ac.uk
[2] Computing At School, UK
simon.humphreys@computingatschool.org.uk

Abstract. After a period of intense activity in preparation for the transition, Computing has been implemented in the curriculum in England for all children from ages 5-16. In this paper we investigate the aspects of professional development that Computing teachers are utilising. We conducted a survey of over 900 Computing teachers in England and use the results to reflect on the benefits of face-to-face vs online communities to support teachers. Our results show that teachers find the face-to-face events and training to be useful, and that teachers in our community are participating in many hours of professional development in order to address their needs in content knowledge and pedagogical content knowledge in Computing. Furthermore an online community is valuable in supporting teachers who require resources, access to expertise and guidance on curriculum issues in addition to face-to-face training, networking and support.

Keywords: Computer science teacher education, teacher professional development, computing education.

1 Introduction

Computing has now been implemented in the curriculum in England for all children from ages 5 -16; the rationale and preparation for this was described in [3,4]. The aims of the new curriculum are that all pupils:

- can understand and apply the fundamental principles and concepts of computer science, including abstraction, logic, algorithms and data representation
- can analyse problems in computational terms, and have repeated practical experience of writing computer programs in order to solve such problems
- can evaluate and apply information technology, including new or unfamiliar technologies, analytically to solve problems
- are responsible, competent, confident and creative users of information and communication technology [11]

© Springer International Publishing Switzerland 2015
A. Brodnik and J. Vahrenhold (Eds.): ISSEP 2015, LNCS 9378, pp. 69–81, 2015.
DOI: 10.1007/978-3-319-25396-1_7

Children at all ages will be learning computational thinking skills, partly through learning computer programming. The curriculum includes the following strands:

- Algorithms and Programming
- Data
- Computers and Social Informatics
- Communication and Networking
- IT and Digital Literacy

The advantage for children of learning computational thinking in school from aged 5 with a gradual introduction to computer programming over the period of their whole schooling means that they will be able to consolidate and extend their understanding of the principles of computing gradually, thus hopefully preventing what Lister describes as "fragile knowledge" [19]. However this gradual introduction to Computing is not possible for teachers, particularly secondary teachers, who have to learn computer programming with some time pressure, at a time when there are already many pressures on teachers in terms of their workload.

This paper describes the results of a survey of over 900 Computing teachers in England which has been compared with a survey of a similar number of Computing teachers last year. The study is primarily focused on the type of professional development (PD) activities teachers find useful and what they value from a professional learning community specifically for Computing in schools. We also reflect on the benefits of face-to-face vs online communities to support the PD of these teachers. The purpose of our research was to identify what PD activities Computing teachers are engaged in and find useful, and whether there is any tendency to prefer online or face-to-face activities in the context of PD.

2   Professional Development of Computing Teachers

The move towards the inclusion of computer science in the school curriculum in many countries has led to concerns about how teachers will manage this change and how sufficient teachers can be found [8,27,24,25]. Teachers have a need for new subject knowledge in computer science, but also importantly, they need to gain confidence in their abilities to teach the new subject [27].

Computing is a domain in which teachers may feel isolated [16] and lack confidence [27,25] or a sense of identity [23]. Professional development (PD) in computer science education for teachers can take a number of forms. Training as the primary or only aspect of PD has been criticised by a number of authors [18,7,17], although subject-knowledge workshops for teachers may be one useful form of Computing PD [13]. In New Zealand, preparing teachers for curriculum change has led to the introduction of 2 to 3 day workshops which are followed up with discussion groups with teachers working in clusters [27], exemplifying a type of collaborative PD [7]. Goode describes the provision of workshops in computer science and pedagogy and notes that these cause teachers to develop their own small networks of support. Morrison et al [22] adopted the originally university-focused Disciplinary Commons approach [14] to be used with school teachers, by providing monthly meetings to discuss issues of teaching and curriculum over a period of a year. One substantial US study into PD across all subjects suggests that key elements are: having ongoing training that is connected to practice; focusing on specific curriculum content; and building of strong relationships between teachers [10], and this is backed up by similar findings in the UK [9]. In addition, the benefits of having frequent contact with a provider was a highlighted by a large-scale synthesis of teachers' professional learning in New Zealand [28]. Work in England which relates to the current study is focusing on a holistic model of CPD [26] including training, mentoring and support with a community of practice, following such recommendations from generic PD research.

3   Communities of Practice - Online and Offline

The community of practice (CoP) has been defined as a group of people who "share a concern or a passion for something they do and learn how to do it better as they interact regularly" [30]. Technology-enabled communities of practice [31] can make effective online learning communities in the domain of education but there is also value in face-to-face interaction [6], not least where people are reticent to join discussions [12] and as such do not fully participate in the online community. Online communities contain an "ecology of resources" [20] and have been shown to have many benefits for teachers' PD [29,21].

Online learning courses are not the same as online communities, and with purely online training there are reported issues with retention [1,2]. Although it has been reported that face-to-face interaction for adult learners has no advantage over blended learning [15], that is likely to be because the face-to-face elements of blended learning allow relationships to be established, in comparison to purely online training opportunities, such as MOOCs in computer science. Online learning courses may be better suited to certain domains: the "getting stuck" element of computer programming means that it becomes easy to give up when it becomes difficult [1]. Learning to be a competent computer programmer is a long, slow process that can be difficult to fit in around the daily demands of a busy timetable. Time out at a session locally can be easier to maintain than an online course, and thus an online community that signposts face-to-face PD becomes an option that has many advantages.

4   The Computing at School Community

Computing At School (CAS) is a grass-roots organisation in the UK which has had a great influence on the emerging changes. CAS exists to provide leadership and strategic guidance to all those involved in Computing education in schools in the UK, with a significant but not exclusive focus on the computer science theme within the wider Computing curriculum [4]. CAS has a particular focus on supporting teachers to deliver the new curriculum in the classroom, with confidence and enthusiasm, through building local communities of practice.

The CAS community meets many of the criteria for a community of practice in such that there is a clearly identifiable domain, knowledge and practice [30] in common for teachers of Computing in a context of curriculum change. The formation of regional hubs where teachers could meet after school, in local CoPs with their peers, to share resources, receive training, try out lesson ideas and discuss pedagogy with each other has been the centre point of all CAS activity. In addition to face-to-face meetings happening all over the country, CAS has an online community site that enables teachers to communicate with one another and find out about face to face events [5]. This site has four features: news, discussions, resources and event listings and is the primary place where face-to-face events are advertised and promoted to teachers. Despite the fact that the online site is seen as the centrepoint to many who join CAS, CAS is built around the principles of local, face-to-face, support for teachers, as exemplified by its supportive PD training programme, built on the concepts of mentoring, peer-to-peer support, cascade of subject knowledge and accessible role models [26]. The CAS site is growing on a daily basis with 18000 members at the time of writing, as can be seen in Figure 1.

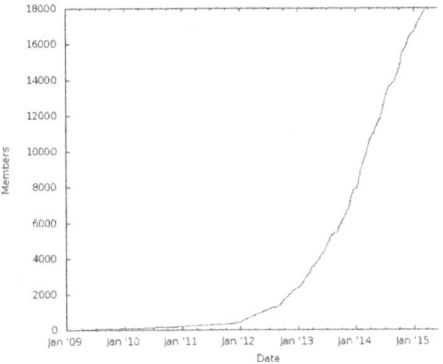

Fig. 1. CAS Membership from 2009-2015

In surveying members of CAS, particularly teachers, the questions our study sought to address are:

1. What are Computing teachers doing to address their PD needs in Computing?
2. How many hours are Computing teachers spending on their PD?
3. To what extent are online and face-to-face activities valued by teachers?

5 Methodology

For the purposes of this study we surveyed a large group of teachers at the beginning of 2014 and then again in 2015 after the curriculum change had taken place. We also have cross-checked our findings against 764 evaluations from the face-to-face training events run locally by experienced teachers with Computing At School. As part of our continual evaluation we also collect data about events held ten weeks after the event . Other aspects of the evaluation process are reported in [26].

The survey was advertised and promoted through the CAS organisation. The vast majority of the 1949 respondents (92%) were members of CAS. For this purpose we have extracted only the responses from teachers in England,which is 981 from the 2015 survey (with 864 from the previous year's survey for comparison). The data were collected using an online tool then extracted into statistical software for further analysis. Teachers gave consent for the data from the surveys they complete to be used to find out more about the community and their engagement with it. Teachers were also asked if they wish to take part in follow-up interviews for more in-depth analysis.

6 Findings

In this section we report on the findings of our survey, contrasted where relevant with the previous year's survey.

6.1 Teacher Profile

In 2015, 65% of the teachers responding teach in secondary education (75% in the 2014 survey), with 31% teaching in a primary (ages 4-11) or middle (ages 7-13) school (21% in the 2014 survey). 4% teach in institutions that only have students aged 16 and over. For the rest of this paper we will ignore this latter group to focus on teachers that teach children affected by the new Computing curriculum. The teachers responding teach different amounts of Computing during an average week (see Figure 2). Since 2014 the number of hours teaching Computing has increased; there is an increase of 7% in the number teaching more than 15 hours of Computing each week and 5% at 10-14 hours per week. This is due to the introduction of the curriculum which was optional up to September 2014. Primary teachers who are mostly generic teachers teach mostly 1-4 hours per week (70%) whereas secondary teachers are more likely to be specialist teachers with 53% at least 10 hours of Computing a week. However there is a small but increasing number of primary and middle teachers who are becoming specialist Computing teachers in their schools.

Teachers were asked how confident they were in their delivery of the Computing curriculum. The mean confidence of a primary teacher (of those answering the survey) was 7.1 and the mean confidence of a secondary teacher answering the survey was 6.8. Figure 3 shows the increase in confidence from 2014 to 2015

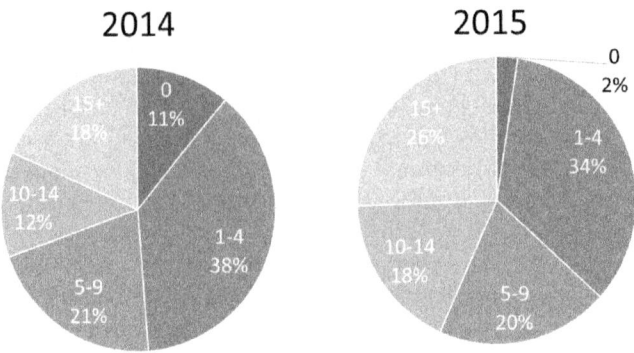

Fig. 2. Hours a week teaching Computing

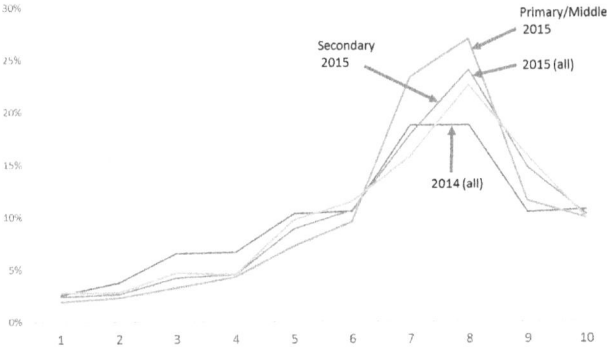

Fig. 3. How confident are you in your ability to teach Computing (1-10)?

with 48% of both secondary and primary/middle teachers reporting confidence of 8 or more.

The next section describes teachers' perception of their PD in Computing.

6.2   Professional Development (PD) in Computing

Teachers were asked which ONE aspect of PD in Computing is most valuable to them. Figure 4 shows that the three most valuable aspects of PD for Computing teachers across our whole sample are:

- Sharing of good practice (26%)
- Attending training events (22%)
- Trying out new ideas in the classroom (20%)

This was then analysed in relation to how confident teachers had previously rated themselves. Teachers rating themselves at least 8 out of 10 are "confident" with teachers rating themselves 7 or less are "less confident". We found that for the less confident teachers more of them identified attending training events (27.5%), followed by sharing of good practice (24.5%) and then being supported by a colleague or Master Teacher (MT) (18%). Both groups of teachers also valued the networking aspect of professional development activities (17.9% for confident teachers and 12.5% less confident teachers saying it was the most valuable aspect for them). A CAS Master Teacher is a teacher who is trained and released from school to support other teachers [26].

Table 1. Teachers/hours on professional development (

No. hours on PD	CAS MT training	University PD	MOOC	Self-Study	CAS Hub
2014					
At least 1 hour	34.2	38.6	35.4	95.4	55.6
More than 6 hours	14.1	20.2	18.7	78.6	13.6
More then16 hours	5.5	11.8	9.3	59.1	2.4
2015					
At least 1 hour	54.2	47.1	39.8	96.3	63.1
More than 6 hours	20.7	28.1	25.7	86.1	17.8
More than16 hours	5.2	17.1	11.6	67.2	3.2

Teachers were asked how many hours they had spent on PD in Computing. Table 1 shows the increase from the 2014 to the 2015 survey. Obviously teachers will on average have spent more time since the previous year's survey but the greatest increase is for the number of teachers who have attended at least one Master Teacher's session (face-to-face); this shows an increase of 20%. There are a number of MOOCs now available for teachers learning computer science subject knowledge - some of these are specifically for teachers. Some teachers are utilising the MOOCs, and we were interesteed to find out whether they had found them useful.

What the survey results show is that 78% of primary/middle teachers and 74% of secondary teachers said that they had found the CAS Master Teacher training useful compared to 52% and 61% of those who attended MOOCs which was a larger difference than in the previous year's survey (see Figure 5). Overall 329 teachers out of 429 attending CAS Master Teacher training (76%) said it was useful or very useful and another 98 saying that parts of it were useful.

Other types of face-to-face PD was also seen as useful with 70% saying that university-provided PD was useful, 70% other CAS events, 68% other non-CAS events and 70% the CAS Conference. Overall 60% teachers said that MOOCs were useful or very useful PD, which was the lowest percentage of all the other types of PD (which were all face-to-face).

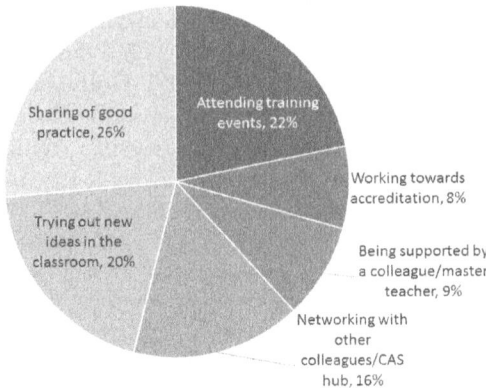

Fig. 4. What type of professional development is most valuable to you?

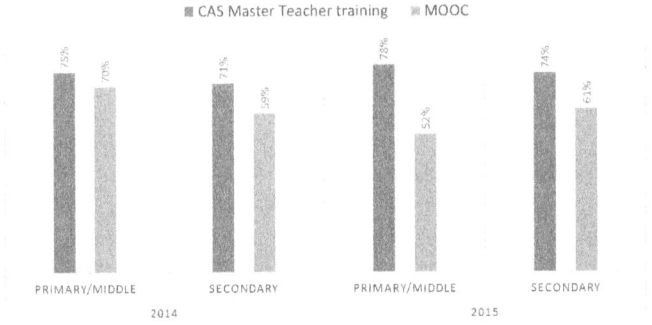

Fig. 5. Type of PD is useful/very useful by year/teacher type (% responses)

6.3   The CAS Community

Teachers were asked which ONE thing was most useful about CAS. The results are shown in Table 2.

Overall the most popular aspect of the CAS community is the sharing of teachers' resources. Teachers voluntarily upload resources that they have developed for their classes for other teachers to share. However, the results differed for different groups of teachers. Primary teachers most valued the guidance on teaching the Computing curriculum (35%), whereas the secondary teachers most valued the access to other teachers' resources (35%). Certainly, overall, the most popular aspect of CAS is the access to resources from those given (33%) . Some teachers gave other valuable aspects:

"Through CAS I have made contacts with other organisations that are helping me improve my ability to teach the computing curriculum"

Table 2. Benefits to teachers of Computing At School

Single most important benefit of CAS (2015)	Primary/Middle	Secondary
Guidance on teaching the Computing curriculum	35%	25%
Access to others' resources	26%	35%
Subject knowledge training	16%	14%
Access to others' experiences	9%	14%
Meeting other supportive colleagues	5%	8%
Other	9%	3%

"Finding out how other people are addressing delivery and assessment of the new computing curriculum"

We compared the confidence of teachers against what they most appreciated about CAS. Less confident teachers were more likely than confident teachers to indicate that the subject knowledge training was most valuable to them (24% compared to 11%).

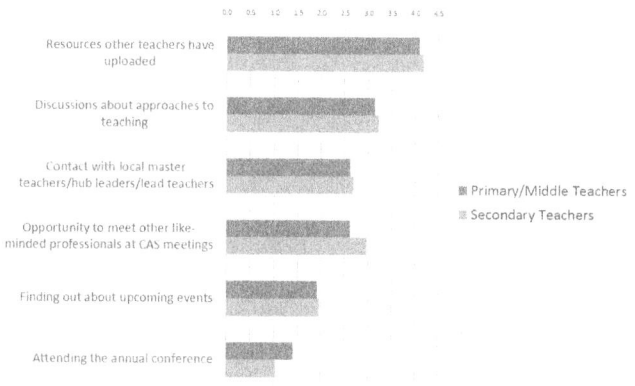

Fig. 6. Features of CAS ranked in order of value

Teachers were also asked to rank the aspects of Computing At School that they valued. Figure 6 shows again that access to other resources are useful, alongside discussions about approaches to teaching, particularly for secondary teachers.

Teachers reported on how often they accessed aspects of the online community. 46% of members viewed the discussion sections of the community site at least weekly (56% in 2014), 26% the events section at least weekly (33% in 2014), and 58% looked at the resources at least weekly (63% in 2014). This indicates that regular accessing of the site has gone down in the previous year (although the number of members has more than doubled).

7   Discussion

The CAS model of PD is built on the belief that face-to-face interaction is the preferred vehicle for supporting subject knowledge development [26]. This is because teachers work in a face-to-face environment by the nature of their role so are comfortable with this type of interaction, and because the potential challenges of learning the subject mean that the confidence building elements of face-to-face training are needed.

The results of this survey has highlighted the following:

- In the year between the two surveys, there has been an increased attendance at face-to-face training.
- Teachers report face-to-face training to be more useful than MOOCs.
- Less confident Computing teachers report that the most valuable aspect of PD is attending training events.
- Confident teachers most value the sharing of good practice and trying out new ideas in the classroom as PD.

This implies that face-to-face learning is important to Computing teachers, although the online community is also important. When taking feedback from events (analysis of 764 forms), 99% of these teachers stated that face-to-face interaction was an important or very important consideration when choosing PD, with 95% also valuing local delivery of training. Darling-Hammond [10] emphasise the importance of strong working relationships between teachers for effective PD and this can be achieved by the kind of face-to-face interactions that are facilitated through CAS.

The support amongst teachers for "trying out ideas in the classroom" also encourages us with our current accreditation programme for teachers that is focused partly around classroom investigations into pedagogical approaches appropriate for teaching Computing[1]. This also relates to research that indicates the importance of relating to practice [10,9].

The fact that teachers are accessing some of the features of the online community less often may be due to the fact that there is, even in one year, more social media available for teachers and an increasing number of websites and organisations supporting Computing. The CAS membership has almost doubled in 12 months, with an ever increasing number of teachers grateful for the resources that teachers freely share amongst themselves. The plethora of online resources mean that it can be time consuming to even locate the appropriate help. Benda et al give examples of students looking for resources posted by others rather than contributing to discussions themselves [1]. Resources on CAS have a higher viewing than discussion items; this seems feasible in a time when teachers are increasingly busy and under pressure in all areas.

As Benda et al aptly describe, programming is hard and finding time to do this online is very difficult [1]. Teachers who need subject knowledge development in addition to the resource-sharing benefits of such a vibrant community need to

---

[1] http://computingatschool.org.uk/certificate

be able to set aside clear blocks of time to do this and this can be more easily achieved with a commitment to a local course. We suggest that the CAS Master Teacher training and support offered within the CAS community provides both the focus on curriculum content recommended by [10] and the close relationship with a provider that is recommended in [28].

8   Conclusion

In this paper we have sought to describe how teachers are accessing and utilising PD in Computing. Our results show that teachers find the face-to-face and locally delivered opportunities very useful, and that teachers in the community are participating in many hours of PD in order to address their needs in content knowledge and pedagogical content knowledge in Computing. Furthermore an online community is valuable in supporting teachers who require resources, access to expertise and guidance on curriculum issues in addition to face-to-face training, networking and support.

References

1. Benda, K., Bruckman, A., Guzdial, M.: When life and learning do not fit: Challenges of workload and communication in introductory computer science online. Trans. Comput. Educ. 12(4), 15:1–15:38 (2012)
2. Boston, W., Diaz, S.R., Gibson, A.M., Ice, P., Richardson, J., Swan, K.: An exploration of the relationship between indicators of the community of inquiry framework and retention in online programs. Journal of Asynchronous Learning Networks 13(3), 67–83 (2009)
3. Brown, N.C.C., Kolling, M., Crick, T., Peyton-Jones, S., Humphreys, S., Sentance, S.: Bringing Computer Science back into Schools: Lessons from the UK. In: Proceedings of the 44th ACM Technical Symposium on Computer science Education, SIGCSE 2013. ACM (2013)
4. Brown, N.C.C., Sentance, S., Crick, T., Humphreys, S.: Restart: The Resurgence of Computer Science in UK Schools. ACM Transactions of Computing Education 14(2) (June 2014)
5. Brown, N.C.C., Kölling, M.: A tale of three sites: Resource and knowledge sharing amongst computer science educators. In: Proceedings of the Ninth Annual International ACM Conference on International Computing Education Research, ICER 2013, pp. 27–34. ACM, New York (2013)
6. Cooper, S., Grover, S., Simon, B.: Building a Virtual Community of Practice for K-12 CS Teachers. Communications of the ACM 57(5), 39–41 (2014)
7. Cordingley, P.: The Impact of Collaborative CPD on Classroom Teaching and Learning: Review: What Do Teacher Impact Data Tell Us about Collaborative CPD? EPPI-Centre, Social Science Research Unit, Institute of Education, University of London (2005)
8. CSTA: Running on Empty. Tech. rep (2010), http://runningonempty.acm.org/
9. CUREE: Understanding what enables high-quality professional learning. Tech. rep. Pearson (2013)

10. Darling-Hammond, L., Wei, R.C., Andree, A., Richardson, N., Orphanos, S.: Professional learning in the learning profession. National Staff Development Council, Washington, DC (2009)
11. Department for Education: National Curriculum for England: Computing programme of study. Tech. rep., Department for Education (2013), https://www.gov.uk/government/publications/national-curriculum-in-england-computing-programmes-of-study/national-curriculum-in-england-computing-programmes-of-study
12. Dron, J., Seidel, C., Litten, G.: Transactional distance in a blended learning environment. Research in Learning Technology 12(2) (2004)
13. Ericson, B., Guzdial, M., Biggers, M.: A Model for Improving Secondary CS Education. SIGCSE Bull. 37(1), 332–336 (2005)
14. Fincher, S., Tenenberg, J.: Warren's question. In: Proceedings of the Third International Workshop on Computing Education Research, ICER 2007, pp. 51–60. ACM, New York (2007)
15. Fishman, B., Konstantopoulos, S., Kubitskey, B.W., Vath, R., Park, G., Johnson, H., Edelson, D.C.: Comparing the impact of online and face-to-face professional development in the context of curriculum implementation. Journal of Teacher Education 64(5), 426–438 (2013)
16. Goode, J.: If you build teachers, will students come? the role of teachers in broadening computer science learning for urban youth. Journal of Educational Computing Research 36(1), 65–88 (2007)
17. Guskey, T.R., Yoon, K.S.: What works in Professional Development? The Leading Edge (2009)
18. Kennedy, A.: Models of continuing professional development: a framework for analysis. Journal of In-Service Education 31(2), 235–250 (2005), http://www.tandfonline.com/doi/abs/10.1080/13674580500200277
19. Lister, R., Adams, E.S., Fitzgerald, S., Fone, W., Hamer, J., Lindholm, M., McCartney, R., Moström, J.E., Sanders, K., Seppälä, O., Simon, B., Thomas, L.: A multi-national study of reading and tracing skills in novice programmers. In: Working Group Reports from ITiCSE on Innovation and Technology in Computer Science Education, ITiCSE-WGR 2004, pp. 119–150. ACM, New York (2004)
20. Luckin, R., Weatherby, K.: Online learning communities in context. International Journal of Web Based Communities 8(4), 440–454 (2012)
21. Matzat, U.: Do blended virtual learning communities enhance teachers' professional development more than purely virtual ones? a large scale empirical comparison. Computers & Education 60(1), 40–51 (2013)
22. Morrison, B.B., Ni, L., Guzdial, M.: Adapting the disciplinary commons model for high school teachers: Improving recruitment, creating community. In: Proceedings of the Ninth Annual International Conference on International Computing Education Research, ICER 2012, pp. 47–54. ACM, New York (2012)
23. Ni, L., Guzdial, M.: Who Am I?: Understanding High School Computer Science teachers' professional identity. In: Proceedings of the 43rd ACM Technical Symposium on Computer Science Education, SIGCSE 2012, pp. 499–504. ACM, New York (2012)
24. Schulte, C., Hornung, M., Sentance, S., Dagiene, V., Jevsikova, T., Thota, N., Eckerdal, A., Peters, A.K.: Computer science at school/cs teacher education: Koli working-group report on cs at school. In: Proceedings of the 12th Koli Calling International Conference on Computing Education Research, Koli Calling 2012, pp. 29–38. ACM, New York (2012)

25. Sentance, S., Dorling, M., McNicol, A.: Computer science in secondary schools in the uk: Ways to empower teachers. In: Diethelm, I., Mittermeir, R.T. (eds.) ISSEP 2013. LNCS, vol. 7780, pp. 15–30. Springer, Heidelberg (2013)

26. Sentance, S., Humphreys, S., Dorling, M.: The Network of Teaching Excellence in Computer Science and Master Teachers. In: WIPSCE 2014 (Workshop in Primary and Secondary Computing Education). ACM (2014)

27. Thompson, D., Bell, T.: Adoption of new Computer Science High School Standards by New Zealand teachers. In: Proceedings of the 44th SIGCSE Technical Symposium on Computer Science Education, SIGCSE 2013. ACM (2013)

28. Timperley, H.S., Parr, J.M., Bertanees, C.: Promoting professional inquiry for improved outcomes for students in New Zealand. Professional Development in Education 35(2), 227–245 (2009)

29. Tseng, F.C., Kuo, F.Y.: A study of social participation and knowledge sharing in the teachers' online professional community of practice. Computers & Education 72, 37–47 (2014)

30. Wenger, E.: Communities of practice: A brief introduction. National Science Foundation (US) (2011)

31. Wenger, E., White, N., Smith, J.D.: Digital habitats: Stewarding technology for communities. CPsquare (2009)

Programming in Scratch Using Inquiry-Based Approach

Jiří Vaníček

University of South Bohemia in České Budějovice, Czech Republic
vanicek@pf.jcu.cz

Abstract. Inquiry-based learning has recently become one of the much advocated methodologies used especially in teaching of natural sciences. This of course opens the question whether it is suitable also for teaching informatics, or even more specifically for teaching programming in a didactical programming environment. In action research conducted within teaching practice of pre-service informatics teachers we tried to explore if any of the approaches and types of activities used by these pre-service teachers correspond to the principles of inquiry-based learning. In the research we study how pre-service teachers cope with these approaches to teaching, what contemporary pupils' attitudes to learning programming are, how this topic and the different activities used in the lessons appeal to them and how this type of lessons is perceived by teachers themselves – pre-service informatics teachers.

Keywords: teaching programming, lower secondary school, inquiry-based learning, Scratch.

1 Inquiry-Based Learning and Programming Education

The world-wide trend of implementing areas from informatics into school curricula and into teaching at lower school levels brought fruit in the Czech Republic last year when the government adopted the Strategy of Digital Education 2020. This strategic document lists three priority objectives, one of which is the development of pupils' informatics thinking [1, p. 14]. Putting this government resolution into practice can be expected to have impact on implementation of informatics curriculum into lower secondary education in a much greater extent than was usual.

For many years, teaching mathematics and natural sciences on primary and secondary schools have been based on inconvenient teaching methods based on mere reproduction of knowledge, on instruction and tutorials that do not result in deeper comprehension and do not develop thinking skills. Inquiry-based learning (IBL) is a response to this situation. It is a method in which a pupil's knowledge is constructed through a system of questions and problem solving [2]. The pupil observes the world around them and based on these observations proposes possible explanations to the solved problem [3]. According to Papáček, a teacher becomes the knowledgeable guide in problem solving and guides the pupil in a way that resembles the progress of work in real research [2]. In correspondence to constructivist theories of learning the pupils are active, solve problems, gain experience and are guided to the construction

© Springer International Publishing Switzerland 2015
A. Brodnik and J. Vahrenhold (Eds.): ISSEP 2015, LNCS 9378, pp. 82–93, 2015.
DOI: 10.1007/978-3-319-25396-1_8

of a cognitive model of the observed phenomena. The pupil's active and creative position in the lesson is a necessary condition for successful learning [4].

Inquiry is the process in which we do not ask "What is it that we know?" but "What are the things that we do not know and what questions can we ask about them?" [5, p. 8] Baptist quotes the American mathematician Paul Halmos whose request was: "Don't preach facts, stimulate acts." [6, p. 7].

IBL approaches can be classified with respect to which part of the inquiry process is undertaken by the teacher. Eastwell distinguishes between four types of inquiries: confirmation, structured, guided and open researches.

— Confirmation research – students are given the question and action (method), the results are known in advance, the purpose is to verify a principle through pupils' own experience.
— Structured research – the teacher presents a question and the possible action (method), based on their knowledge pupils determine the explanation of the studied phenomenon.
— Guided research – the teacher presents a research question but leaves the method and solution up to the pupils.
— Open research – pupils present a question, think of a method, conduct the research and define the results [7].

The question to answer is how to employ inquiry-based approaches in teaching informatics, namely in teaching programming (and to what extent the environment of informatics can be used to teach a pupil in an inquiry-based way). How difficult is it to implement the different stages of inquiry-based learning in teaching programming without actually having to change the structure of the subject and the priorities of the teaching goals? What elements should be accentuated and what tools should be used to ensure that the learning process includes the main inquiry processes according to Bell et al. [3]: Orienting and asking questions, Hypothesis generation, Planning, Investigation, Analysis and interpretation, Model, Conclusion and evaluation, Communication and Prediction?

One of the significant learning theories that have influence on how programming is taught is Papert's constructionism. The theory of constructionism is based on the belief that learning is more a process of (re)construction than transmission of knowledge and opposes the so called instructionalism. According to this theory a learning process is more efficient if it has the form of active construction of meaningful artefacts. "A major constructionist method is to mix media in model construction and to translate from one media, say, a mathematical function, into another media, say, words or diagrams." [8]. Papert emphasises a pupil's verbal communication and also the tools a pupil creates and manipulates with while constructing knowledge. In the communication the pupil formulates ideas for the computer. Papert uses the term microworld to refer to the subset of reality (or constructed reality) whose structure corresponds to the given cognitive mechanisms of the learner, which provides an environment that can work efficiently and allow the learner to practice selected substantial ideas or mental skills [9].

1.1   Teaching Programming

Many Logo-like educational programming environments such as Scratch, NetLogo, Imagine etc. are tools that support hypothesis generation, prediction, model, evaluation. It can be presumed that a well-planned and conducted lesson in these environments based on constructionist methodologies will support pupils' discovery and experimenting, will guide them to formulation and verification of hypotheses, to their evaluation and will help them discover causes through analyses of the behaviour of the system.   However, can these partial activities be put together to make one organic whole turning "the pupil into a young inquirer"? Is informatics equally as rich an environment for IBL as other natural sciences? A microworld with its low floor, high ceilings and wide walls [10] can be expected to give pupils answers to questions that have not even been worded but that can be tested in experiments if pupils learn to be perceptive to these answers. Do didactical programming environments allow the pupils to conduct "real" experiments similar to experiments in biology, physics and chemistry with authentic artefacts and natural phenomena? Will pupils take the "artificial" computer environment where reality is simulated as sufficiently real to be willing to inquire into it in the same way they would do in natural sciences? Can the topic of programming support the orientation of IBL on the development of a learner with scientific thinking without distracting attention from the main goal of the topic itself, i.e. to develop a learner who has algorithmic thinking, who can grasp and solve a problem, model a situation and then implement the model to a computer?

These questions are very difficult to answer. Bearing this in mind our goal was to modify the standard teaching of programming to 15-year-old pupils by including some IBL elements (some of these overlap with constructionist approaches) with main objective laid in teaching programming.

We compared approaches to teaching programming used in Czech, Slovak and several other textbooks and also methodological manuals for teaching programming available on the internet [11]. Predominant in these textbooks and manuals are two basic types of programming tasks:

— short programming tasks (etudes), whose goal is to practice one particular skill or item of knowledge
— larger programming units (projects) whose goal is to create some larger work, i.e. a computer game

Teaching through a series of several-minute long programming etudes always targets at acquisition of one specific skill, is oriented on one specific item of knowledge. It is often the case that the difficulty of these etudes grows gradually and results in a deeper understanding of a concept. Motivation of these activities is very often weaker than in case of larger projects. Pupils can easily check their result but the tasks do not develop their creativity.

Teaching through larger programming units – the so called "projects", e.g. creating a game, often has the form of a sequence of activities organized as a tutorial or a problem task. The reason why the large project is divided into the subtasks is the teacher's struggle to ensure the extensive activity is concluded successfully and in due time. It is hard to combine the goal to teach a specific concept, procedure or method with open-ended activities. These activities are highly motivating for pupils, are more authentic but cannot target the development of one particular skill.

In the two ways of selecting types of problems, two different approaches to teaching goals can be observed: in case of shorter programming etudes the approach is intensive, focused on competences in the area of concepts, closer to traditional teaching of mathematics, and the other is holistic, more open, emphasising creativity, focusing on product. We are fully aware that the term project for larger programming units is not accurate as these projects are more a sequence of tasks leading to the creation of a product, a program, which is in some cases, e.g. in Imagine, called a project.

2   Project and Methodology

The descriptive case study [12] of teaching programming enriched by elements of inquiry-based approach was conducted on one semester project of programming in the environment Scratch [13] at a town lower secondary school in autumn 2014. The teaching experiment was conducted with a group of 13 15-years-old pupils from an ordinary lower secondary school, with an above average number of ICT lessons in its School Education Programme (1 lesson a week in all the previous 3 grades). The curriculum in the previous grades included no informatics, its focus was on user approaches (e.g. Office, Internet and presentations, editing photographs, vector graphics). The pupils were taught in traditional teaching schemas and had no former experience with IBL elements in the area.

The teaching in the project was conducted by pre-service informatics teachers in their final year of master studies. It was conducted as their compulsory teaching practice. The pre-service teachers took turns in teaching the lessons. The lesson plans were prepared under supervision of the educator in charge of this teaching practice. The lessons were planned more or less according to a syllabus created by comparing the available teaching manuals for work with children programming environments Scratch and Imagine. Attention was paid to inclusion of new programming structures, elements and tools. In the initial lessons the pupils got acquainted with the environment. Then they were introduced to the properties of objects (e.g. costumes), cycles and decisions, events and running actions, communication among objects, synchronization among threads, number variables. This took 15 lessons. The tasks and problems were taken from the internet Scratch Store and if they were modified, the changes usually affected graphics or the story told in the activity.

The types of IBL used most frequently were confirmation and structured research. In order to follow the principles of inquiry-based learning the teachers tried to plan and conduct their lessons in such a way that they and their pupils would be working on more complex problems allowing chaining to simpler tasks, creation of situations in which it was possible to ask questions. These were for example simple games or stories of characters. The teachers first had to divide these complex projects into partial subtasks that were then given to the pupils for solution. Thus the pupils were guided through a series of shorter tasks with intrinsic motivation to achieve the goal of creating some more extensive work. The pupils were given the chance to experiment but at the same time check the partial results and if needed to copy the correct solution and thus could continue with solution of the next subtask.

The teaching experiment also included a pupils' programming project Christmas greeting. In the project it was entirely up to the pupils what the content would be. Most of the work on the project was done outside of school. However, the key parts of the project – introduction, conclusion and consultations on the project work took place at school.

In this case study, the data were collected by the method of participant observation and subsequent joint reflection on the lessons. This allowed us to observe both the pupils' attitudes and the pre-service teachers' activity, in which tasks they were using the selected IBL teaching strategy and how they were gradually building the syllabus. Each lesson was followed by 1 hour of joint reflection on the observed lesson and 2 hours of planning of the following lesson with the help of the supervisor and the class teacher. Another method of data collection was analysis of pupils' work. In this analysis we tried to discover not only how the pupils were proceeding but also how their work was assessed by the pre-service teachers and what they were looking for in the pupils' work. Upon termination of the teaching experiment, a questionnaire was administered to the pupils and structured interviews with were conducted the participating pre-service teachers.

3   Findings

This pilot project of teaching programming in didactical programming environments with corresponding methodologies involves a number of elements of inquiry-based learning on the condition that pupils are provided with enough space to communicate and to solve (sub)tasks and problems. The major difference is in the goal of teaching – in IBL the goal is inquiry-based approach resulting in acquiring new knowledge and being able to apply it, whereas in programming the goal is making use of this type of thinking or knowledge for making a product, i.e. a computer programme.

The idea on the background of this experiment was that in-service teachers who would be beginners in teaching programming would act similarly to the participating pre-service teachers. We studied what types of problems and tasks pre-service teachers were selecting for their lessons and how successful they were in the lesson planning process (we did not assess how successfully the lesson was then conducted). The participating pre-service teachers found it easier to plan a lesson as a sequence of programming etudes and to make a whole lesson plan from these activities. When teaching programming through more extensive activities ("projects") the teachers did not even expect their pupils to be able to divide the complex problem into subtasks and did this division for them. They were aware that this was hard work even for them as teachers.

3.1   Teachers' Approaches and Mistakes

In the following part we present comments on some of the mistakes made by the pre-service teachers.

Preference of Creation of Projects. The pre-service teachers often planned lessons based on creation of a larger project; they found it more motivating for their pupils and may be even for themselves. This proved to be problematic, especially as it turned out that the projects very often required some knowledge the pupils did not have and which was impossible to be taught because of time constraints and because of the whole conception of the teaching. The teachers often solved this by presenting the most difficult passage to the pupils in the form of a created code or by asking them to copy it without actually clarifying the meaning (for example using cloning in the currently programmed game). Time management proved to be very problematic; it often happened that the project activities were not concluded within one lesson and had to be finished in the following one. This disturbed the conception of the teaching experiment; it had then to be finished by another pre-service teacher who, in consequence, had less time to conduct their own lesson.

We presume the activity of a teacher who would be a beginner in teaching programming, who would not have the needed methodological background and would want their pupils to create a programmed product would be similar to the activities of the pre-service teachers. The teacher's ambition to "achieve something great" with their pupils has effects on fulfilling the teaching goals; it can result   in non-conceptual teaching, in insufficient practice in the basics which may later result in a lack of ability to program simpler tasks.

Too Long Code. Another frequent mistake in the lesson planning was that the teachers proceeded to long multiline codes too quickly and the pupils found it hard to orient themselves in them. The teachers failed to see how to avoid these multiline codes, e.g. by using more threads, communication among objects etc. Their own knowledge of programming in different languages and the methodological course of programming in Scratch did not help them overcome this problem.

Endless Loop. A typical mistake whose origin is in the environment Scratch itself is the use of endless loop. Scratch offers this as a solution of simple programming tasks for beginners. Activities using this structure are motivating and can make pupils attracted to programming. However, both teachers and pupils found it difficult to get rid of this bad habit at later stages and instead of using a simple cycle with a condition they were creating endless loops with nested decision blocks and hard terminations of the running code.

Bad Synchronization among Objects. Another problem that can happen in simple object programming is bad synchronization which the programmer (and also the inexperienced teacher) fails to see. It becomes a source of problems the programmer (and the teacher) is not able to find. They tend to blame the application for the mistake. Let us illustrate this on an example we came across in our observations. In the "cat and mouse game" the pupils defined that if the mouse bumped into the cat, 1 point would be deduced from the score, the cat would meow and the mouse would jump to the beginning of the page (fig. 3 on the left). The initial score was set as the number of mice to be caught. However, bad synchronization of the commands for sending and receiving the message with the command for the mouse to jump to the beginning of the page resulted in a state when, before the cat finished saying "Meow", the program detected repeatedly that the mouse was touching the cat and made several deductions from the score. Thus the score had been nulled already when the first mouse was caught. This type of mistake in the code is very hard to detect for a beginner teacher. The participating teachers did not manage to detect it in the situation.

Approaches to Pupils' Assessment. The teachers were also expected to assess pupils' work. It was very interesting to observe how they do it. In general the teachers were checking that the programme was running correctly. In case of projects they were also sometimes assessing which of the taught programming elements and techniques the pupils had used in their work (they tried to evaluate quantitatively whether the pupils were using a cycle, message, decision, costumes etc.). In their assessment they did not take into account whether a pupil used some original technique or approach or a witty solution by using some programming element in a non-traditional way (e.g. removing matches from the table can be solved visually by hiding the matches by interaction with each object of a match, or by changing a costume of a heap of matches as one object, or by gradual covering of a matches in a row by another object). The theme of the work or original ideas were not part of the assessment. And if the teachers found the theme not meeting their idea of decorum, they did not allow the author to present it among other successful projects even if it fulfilled all the other criteria.

3.2   Pupils' Reactions

Initial Hesitation and Very Slow Progress. In the first lesson the pupils were rather hesitant and did not seem to be very enthusiastic about the idea of programming. This was largely due to the fact that they were not able to orient themselves in the environment and the type of mental activity that would be required from them. However, about one third of them got really excited about programming thanks to the idea of creating stories. These pupils were happy to work regularly on voluntary homework. In consequence they were more confident during lessons, were more active and happy to work on their own.

No significant progress in pupils' knowledge could be observed after the first 8 lessons in the first 2 months. With a few exceptions the pupils were not able to solve problems and tasks on their own, they did not remember the commands and techniques of work, they were not able to discover which command or technique to use in a given situation. Modification of work set for homework had usually only the form of a change in the theme or background or costumes, but not of the script or scenario. However, the pupils showed they were able to solve simple programming tasks during an inserted lesson "Hour of Code" in which they were programming a set of tasks of turtle graphics in the environment of the film Frozen [14]. Still we could observe certain helplessness of weaker pupils if they had arranged a sequence of commands into the body of the cycle inappropriately and could not find a way to get rid of the wrong script. They usually turned to the trial and error method without even trying to be systematic (e.g. trying out all possibilities).

Our participant observation made us conclude that the pupils were not given sufficiently varied programming tasks. They were given tasks based on creation of a code, not on detection of a wrong code in which they would have to be looking for the cause of a mistake. Also much more time would be needed for the pupils to master even the very basics of programming.

Pupils' Project. The most significant progress in the pupils' knowledge could be observed after having worked on a short project of creating an electronic Christmas greeting with a theme of their own choice. The pupils had to take responsibility for finishing the work, which might have been the main factor causing that after this project was over and after a 3 week break the pupils grew more perceptive to explanations, more able to work on their own and to justify. Still, with a few exceptions projects did not bring any original ideas (neither thematic, nor programming).

About one third of the pupils became interested in programming at the very beginning of the teaching experiment. These pupils worked on programming also outside of the classroom and were able to propose original ideas that they could make use of in the lessons by varying the themes of the programming projects (e.g. the cat and mouse game was modified into catching a football ball – Fig. 3 on the right). This group of pupils realized that the nature of very different everyday activities had the same structure in programming and could be modelled in the same way.

Fig. 1. The same game in two attires: a cat is chasing a mouse and football penalties. On the left the taught version, on the right a pupil's spontaneous modification.

Activities Appealing to Pupils. It turned out that successful were those activities in which the pupils were not programming but creating scenarios and dialogues for characters, in which they were drawing costumes, i.e. in which they were modifying the story as a user but not modifying the programme itself as a programmer. If any change in the code occurred, these were usually simple commands run by events, linear stories, there were no cycles and decisions.

We tried to pinpoint which themes in tasks were most favourable for making pupils gain new knowledge:

— drawing of different costumes of the characters: modifying the theme of the activity but modifying the programming code as little as possible
— variables: if they wanted to watch the score of the game
— conditions: if they wanted to finish a game
— messages: if they wanted to direct the dialogue between two or more characters

The pupils preferred making programmes with no cycles (there was a strong preference for simply run actions in more threads or for endless loops – i.e. preference for simple code). It is hard to say whether this is in consequence of unpreparedness of Scratch to use the built-in loop counter (as is *repcount* in Imagine). This could only be answered if the real cycle FOR was needed, which was not the case in the taught lessons.

3.3   IBL Activity: Programming a Game Played Against the Computer

The whole teaching experiment was concluded by an inquiry-based learning activity in which the pupils were asked to program a game where the player would be playing the computer. At first the pupils programmed a game for two players. Then they were asked to discover the winning strategy and then implement the discovered rules into a programme for the computer. For these ends we selected a simple board game Nim in subtraction misére variant with one heap of the size of 13 based on removing matches from the board [15]. The winning strategy in these games is based on reaching the so called winning position (the strategy is described e.g. in [16]. The game offers a simple winning strategy and the pupils can program one player in such a way that he cannot be beaten.

Another goal of this activity was to make pupils realize that the "intelligent computer" must be programmed, i.e. that it is up to the pupils to think out the possible moves for the computer and then program it. This means the computer was to be de-mythologized through this activity (according to Schubert and Schwill one of the important objectives of education is to show to pupils that a computer is only a machine that executes commands [17]).

The basic benefit of this activity for pupils was the search for the optimal winning strategy. The pupils played each other in a game they had programmed themselves and while trying to beat the opponent they were looking for the winning strategy, asking questions, discussing, making conjectures and verifying them. The winning strategy was discovered by 3 pupils within 15 minutes but it took them much longer to realize that if there was an optimal strategy for one of the players, the opponent could never win. They were trying it again and again.

Later, when the winning strategy was grasped by about one half of the pupils after having discussed it with each other and trying it out in matches, the teacher encouraged the pupils to formulate how the computer should play (e.g. "it must always remove matches to four") and to discover how to express this. The subsequent implementation into a code showed how similar the moves of the player and the computer were (Fig. 2). Having finished the activity the best pupils wanted to find out how to modify the game for a different number of matches.

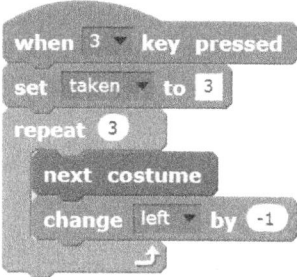 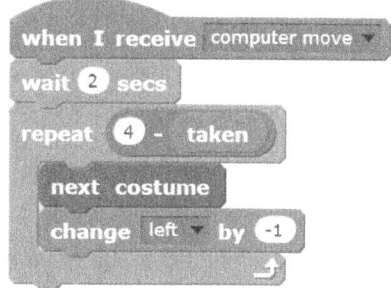

Fig. 2. Analogy between a player's turn and a turn of a computer using an expression in the Nim game

4   Pupils' Perception of the Teaching Experiment

After the teaching experiment the pupils were given a questionnaire with both multiple choice and open items which was meant to allow us to learn more about how they felt and what they thought about programming.

Attitude towards Programming. 80 % of the pupils stated they were good at programming. 20 % did not think it was of any benefit for them to have learned programming unlike pupils in other schools. This negative attitude, however, did not correlate with their success rate because all the pupils who stated they were not good at programming claimed programming was beneficial for them. 80 % of the pupils claimed they would like to continue programming.

We were curious how the course in programming changed the pupils' attitude to their prospective career. 30 % of the pupils explicitly stated they did not want to do any programming in their future career. However, all of these pupils were good both at mathematics and programming.

Teaching Method. The pupils were not able to evaluate the teaching method. We can only guess what teaching method was popular among pupils from which activities the pupils selected as their favourite. The pupils most enjoyed activities of creating a game strategy and implementing it to the computer using IBL, and creation of games. We did not get an answer to the question whether the pupils preferred teaching through miniprojects or through a sequence of programming etudes.

Difficulty of Learning to **Program.** A part of the pupils found it most difficult on programming that it required them to think ("what I found most difficult was when we were assigned a task and we were meant to solve it on our own", "to realize what the thing should do in the programme and how to connect it in a logical way"), part of the pupils spoke of the initial difficulties to get accustomed to the environment and style of work ("before I got acquainted to it"). What they found most interesting on programming was that "they could discover new things", "find out how things work in the computer" and also appreciated the pleasure of the author of an attractive piece of work (a computer game).

The pupils' answers also clearly showed that after the teaching experiment they were still not able to tell what was and what was not programming. Some of the pupils claimed they had had former experience with programming (setting parameters in utilities controlling behaviour of OS or setting animations in 3D graphics).

5   Conclusion

The project of teaching the basics of programming enriched by IBL elements is close to constructionism but has its limits.

5.1   Pupils

It took relatively long before the pupils were able to answer questions and to make any inquiry. We are convinced that this was not caused by the novelty of the unknown environment, as the pupils came across similar situations in different topics, but by their brand new role in their relationship to the computer which they had to get used to (e.g. when a programme was not working, the pupils often asked for the button Undo).

It turned out that the pupils needed a lot of space and time to get used to the ways of a programmer. At that period of time they needed to work with very simple code with simple commands that they were able to vary in a non-programmer way (creation of graphics, search for stories and themes, but no changes in the code). As long as the pupils tend to modify ready-made programmes in this way, it is probably too early to proceed to more difficult parts of programming.

Our observations show that the selected approach of orientation on gradual creation of programming projects in stages does not make pupils want to program, i.e. to write a code and to use control structures. However, it makes the pupils see how a computer works.

5.2   Teachers

The experiment also showed that (pre-service) teachers feel much less confident in teaching programming than in teaching how to use a computer, e.g. use of office applications. This lack of confidence was not so apparent when conducting the planned lesson but in situations when they were asked for help by a pupil whose programme did not work. It seems that detection of a mistake is a very advanced competence. Therefore tasks and problems of this nature must be included in teaching the unit. At the same time activities in which pre-service teachers learn to detect a pupil's mistake must become integral part of courses of didactics of programming in their undergraduate studies.

Acknowledgment. The research was supported by the project GAJU 017/2013/S.

References

1. MŠMT: Strategie digitálního vzdělávání (Strategy of Digital Education). Ministry of Education, Praha (2014), http://www.msmt.cz/file/34429
2. Papáček, M.: Limity a šance zavádění badatelsky orientovaného vyučování přírodopisu a biologie v České republice (Limits and Chance of Implementation of Inquiry-based Learning of Biology in the Czech Republic). In: Papáček, M. (ed.) Didaktika Biologie v České Republice a Badatelsky Orientované Vyučování (DiBi 2010), pp. 129–135. Jihočeská univerzita, České Budějovice (2010),
   https://www.pf.jcu.cz/stru/katedry/bi/
3. Bell, T., Urhahne, D., Schanze, S., Ploetzner, R.: Collaborative inquiry learning: Models, tools, and challenges [online]. International Journal of Science Education 32(3), 349–377 (2010), http://collablitreview.wikispaces.com/
   file/view/collab+inquiry+learning.pdf
4. Hajduković Jandrić, G., Obadović, D.Ž., Stojanović, M., Rančić, I.: Impacts of the Implementation of the Problem-based Learning in Teaching Physics in Primary Schools. The New Educational Review 25(3), 194–204 (2011),
   http://www.educationalrev.us.edu.pl/vol/tner_3_2011.pdf
5. Baptist, P.: Towards new teaching in mathematics. In: Baptist, P., Raab, D. (eds.) Implementing Inquiry in Mathematics Education, pp. 1–12. Universität Bayreuth, Bayreuth (2012) ISBN 978-3-00-040752-9
6. Baptist, P.: Simplify mathematics education. Towards New Teaching in Mathematics 7 (2011) ISSN 2192-7596
7. Eastwell, P.: Letters: Inquiry Learning: Elements of Confusion and Frustration. The American Biology Teacher 71(5), 263–264 (2009)
8. Clayson, J.: Constructionist approaches to creative learning, thinking and education: lessons for the 21st century. AUP Magazine, Paris (Fall 2010),
   http://alumnionline.aup.edu/page.aspx?pid=666
9. Papert, S.: Mindstorms: children, computers, and powerful ideas. Basic Books, New York (1980)
10. Guzdial, M.: Programming environments for novices. In: Fincher, S., Petre, M. (eds.) Computer Science Education Research, pp. 127–154. Taylor & Francis, Abingdon (2004)
11. Krejsa, J.: Výuka základů programování v prostředí Scratch (Education of Basic Programming in Scratch environment). University of South Bohemia, České Budějovice (2014), http://theses.cz/id/b5f11x
12. Yin, R.K.: Case Study Research. Design and Methods, 5th edn. Sage Publication, Los Angeles (2014)
13. Resnick, M., et al.: Scratch: Programming for all. Communications of the ACM 52(11) (2009),
    http://web.media.mit.edu/~mres/papers/Scratch-CACM-final.pdf
14. Code.org: Hour of code. Frozen. Programming online course. Code.org (2014),
    http://studio.code.org/s/frozen
15. Köller, J.: Nim Game. Mathematische Basteleien (2000),
    http://www.mathematische-basteleien.de/nimgame.html
16. Burján, V., Burjánová, L.: Matematické hry (Mathematical games). Pytagoras, Bratislava (1991)
17. Schubert, S., Schwill, A.: Didaktik der Informatik. Spektrum Akademischer Verlag, Heidelberg (2011)

Olympiad in Computer Science and Discrete Mathematics

Athit Maytarattanakhon, Vasiliy Akimushkin, and Sergei Pozdniakov

Saint Petersburg Electrotechnical University "LETI"
5, Professora Popova str., Saint Petersburg, Russia
root@post.etu.spb.ru
http://eltech.ru

Abstract. Many ideas of theoretical computer science is not yet included in the practice of school teaching. To test the methods of learning new ideas one can use the format of school Olympiads which form a
circle of ideas and objectives which can be included in the future general
curriculum.

The paper describes the experience of the Olympiad on theoretical
computer science and discrete mathematics. The Olympiad consist of
two rounds. The first round is held in a distant form but the second one
is held on the premises of universities. All the rounds are organized in an
electronic format and all the participants work with same manipulators
which simulates important concepts or ideas of subject area. Thus, to
the last round of Olympiad, all participants already will be acquainted
with new ideas of subject area and during the time limit can solve more
difficult problems.

As examples we discuss here tasks of DM&TI-2015. They are based on
five manipulators: Turing machines, regular expressions, graphs, Tarski
worlds (predicates and quantifiers) and logic circuits. The paper suggest
a technics for problems design and using of manipulators for solving
problems in computer science and discrete mathematics and technology
for semiautomatic processing of results. The Olympiad uses web services
that provide users feedback and interaction of authors and participants
with problems during preparing and holding of the Olympiad.

Keywords: olympiad, computer science, discrete mathematics, electronic
manipulator, CS competition.

1 Introduction

The importance of theoretical computer science ideas is recognized by the pedagogical society [1]. In Russia the importance of theoretical computer science in
2014 was supported by institutional decision to combine the mathematics and
computer science in one subject area [2,3].

Technics of prevenient introduction of new ideas through Olympiads has own
history in Russia. For example The First URSS Math Olympiad for Vocational
Schools, which was held in 1980, included two extra tours, which can be considered as a bridge between computer science and mathematics:

© Springer International Publishing Switzerland 2015
A. Brodnik and J. Vahrenhold (Eds.): ISSEP 2015, LNCS 9378, pp. 94–105, 2015.
DOI: 10.1007/978-3-319-25396-1_9

1. there was introduced an experimental tour to do simple research work with physical phenomena model on a programmable calculator [4];
2. one of proposed problems required preintroduction of new ideas, so before the Olympiads the popular lection for participants was given, which expand the range of permissible problems [5].

The idea of using new knowledge from an area of computer science is actively used in the Bebras competition [6] and "Construct, Test, Explore" contest (CTE) [7,8]. Also a view on computer science as experimental area is actively develop [9], so an idea of using electronic manipulators to support education in computer science are important too [10,11]. Automation of testing problems solutions is widespread in the programming contests and now they held entirely in electronic form [12].

Using of electronic manipulators to support problem's solution in theoretical computer science can be different: 1) visualizers allow to get acquainted with a new idea, what provide the context for problems statements understanding; 2) local manipulators which don't fix student's action, and so their actions will not be considered as part of the solution (manipulators of such type are used in the competition Bebras, as they do not require changing of interface for answer input and give possibility to use traditional type of input - select one of four variants); 3) finally, there can be such an environments to support search of solution and allow to analyze all the actions of the participant [13].

The Olympiad in theoretical computer science and discrete mathematics is the next step in creationof Internet School in Theoretical Computer Science and Discrete Mathematics at the Faculty of Computer Technology and informatics St. Petersburg Electrotechnical University "LETI" (SPBGTU "LETI"). At the first stage there was an implementation of basic algorithms on graphs manipulators and providing students opportunities to work in three modes: 1) if student choose demonstration mode a new graph will be generated and user get an opportunity to "sweep" algorithm step by step with the comment to each step; 2) in tutoring mode student has the opportunity to "go" algorithm step by step, the system does not allow to do the wrong moves and gives corrective comments 3) in testing mode the system does not check the correctness of each student step, and checks the entire solution [14].

2   Analysis of CS Competition Organization

Competition Bebras passes during a short time (one lesson), usually in the classroom under guidance of teacher in online mode. Preparation to the Bebras competition is based on the work with the tasks of previous years using the comments to the tasks appearing after finishing of competition.

Competition CTE passes during the week in offline mode. Generally the competition begins in the classroom, where students begin to work with the research subjects under the guidance of teachers. After 1–3 hours students were charged interim solution, and then continue to work on the tasks at home, periodically downloading files of improved solutions on the competition website. At the end of competition week teacher gather students to make sure that no one forgot to download the latest versions of solutions.

Programming team competition passes in several online elimination rounds. During the first part of the allotted time after loading solution on Olympiad site and its verification, an information about the verification results (is the solution accepted or not) is disclose to all participants of the Olympiad, which thus can monitor success of rivals and estimate the complexity of the remaining tasks according to the results of other teams. The second part of the allotted time passes after the "freezing" of results visualization, then every team does not know successes of other teams in solving problems and get responses from the system only with verifications their own solutions.

Olympiad in theoretical computer science and discrete mathematics SPBGETU "LETI" pass through three stages. All steps involve the use of same manipulators. As to manipulators of season 2014–2015 there were used the following tools: Turing machines, regular expressions, graphs, the world of Tarski (quantifiers and predicates), logic circuits. The first stage of Olympiad is preparatory tour. At this stage, participants work with the preparatory tasks and get tips on interim action. Time to solve training problems are not limited. After the end of the tour, participants get an analysis of tasks's solutions. The second stage is a extramural round which restricted by 3 hours. The round passes in online mode. Problems is included in this round are not synonymous with training tasks but are formulated in terms of the same manipulators, which already known by participants. The last the third stage is internal round of the Olympiad. On this round participants are invited to come to universities for last Olympiad tour. They get the same amount of tasks and the same amount of time as on previous stage. At the same time, tasks are substantially different from tasks of the previous stages, therefore the preliminary stages are not prepare them to solve "such types" tasks, but give acquaintance with manipulators and new theoretical concepts which will be used in tasks. This give possibility to limit time of Olympiad by 3 hours.

3   Analysis of DM&TI-2015 Content

Tasks described below are based on the same concepts and manipulators that problems for previous rounds. Therefore participants already got acquaintance with concepts such as graph connectivity, planarity, graphs equality (isomorphism), regular expression, a Turing machine, quantifiers, and so on.

Participants are encouraged to first solve the problem without "stars". They get up to 3 points after solution of every such task. Other tasks ("with stars") can give up to 6 points. These tasks areconsidered as additional, that is, the participant can solve those of them which are understandable and interesting for him or her.

1. Graphs. The tasks is accompanied with manipulator, in which one can construct graph and move its vertices to explore the graph on planarity (the manipulator is constructed so that when moving the vertices they not coincided).

The presence of the manipulator saves students from reading of formal math definition for connectivity and planarity (although there is a "help" button inside every task).

1.1. There is a connected graph with 9 vertices, in each vertex converge two edges (the degree of each vertex is equal to 2) and edges do not intersect. Add to the graph as many new edges as possible so that the degree of all the vertices remain equal, and graphs remains planar (i.e, the vertices can be moved so that the edges do not intersect).

1.2*. Let $f(n)$ be the maximal vertex degree of a connected regular planar graph with n vertices (the regularity means that degrees of all vertices are equal). Find and justify upper and lower bounds for $f(n)$.

See solution for this task in the next section.

2. Combinatorics. *There is manipulator attached, which allow to build graphs. The concept of isomorphic graphs is formulated in "terms of the manipulator" and can easily be checked in practice.*

2.1. There is a connected graph with 6 vertices and with exactly one 3-cycle. Construct all different connected graphs with 6 vertices ("different" means non-isomorphic graphs which can not be converted one into other by dragging vertices) and with exactly one cycle (the cycle contain 3 vertices)?

2.2*. How many different (nonisomorphic) connected graphs having n vertices, and exactly one cycle comprising $n-3$ vertices exist?

3. Regular expressions + combinatorics. *On the problem of the manipulator is attached, checks belonging to set of objects described by a regular expression.*

3.1. Construct a regular expression that describes correct formula in the mathematical sense, which comprised numbers "2", two operations (+ and *) and the square brackets. It is prohibited to place ones brackets inside other brackets.

Examples of correct formulas: 2; 2+2*2; [2+2+2]*2+[2+2]; 2+2*[2+2]+2; [2]

Examples of incorrect formulas: 22; [2 * [2 + 2]]

Solution. One possible variant: $(2(+|*)|([(2(+|*))^\wedge 2](+|*))^\wedge (2|[(2(+|*))^\wedge 2])$

3.2. How many formulas satisfying these rules contain exactly 11 characters (you must count all different expressions, even if they give same results after evaluation)?

4. Logic circuits. *The task is based on the manipulator, in which student can do logic circuit from the logic elements AND, OR, NOT (each signal can be sent to the inputs of several elements).*

4.1. Construct a logic circuit which take as inputs two figures a and b of binary number ab output three figures c, d, e of binary number cde= ab +1.

Solution. $e =$ NOT b; $d =$ (NOT b AND a) OR (b AND NOT c); $c = a$ AND b

4.2*. Find upper bounds for quantity of elements of the scheme which add 1 to a binary number with n figures.

Solution. Let $a(n)a(n-1).....a(0) + 1 = b(n+1)b(n)....b(0)$ $p(0) = 1$; $b(k) = a(k) + p(k) =$ (NOT $a(k)$ AND $p(k)$) OR ($a(k)$ AND NOT $p(k)$); $p(k+1) = a(k)$ AND $p(k)$

Answer: $1 + (n-1)*5 + 1 = 5n - 3$

5. Definition of mathematical algorithm: the Turing machine. *The task is accompanied by amanipulator, which simulate Turing machine over the alphabet 0; 1; a; b; , where symbol denotes "Empty" symbol.*

5.1. Turing machine T1 copies string of binary digits by placing copy to the right and separating it by symbol * from initial string (symbol * initially fill all free tape cells). The machine head at the beginning and at the end of the algorithm execution indicates the first unit from the left. Example, the string 111011 will be converted to the string 111011*111011.

Machine T2 subtract numbers in a unary system, herewith the second number from the left is less or equal to the first number. The machine head at the beginning and at the end of the algorithm execution indicates the first unit from the left. Example, the string 11111*11 will be converted to the string 111.

T3 machine multiplies numbers in unary notation. Example, the string 111*11 will be converted to the string 111111.

All mentioned above Turing machines use auxiliary symbols a and b.

Combining these machines, construct a Turing machine T, which performs operation $A^2 - B^2$ for two numbers $(A > B)$ in an unary notation.

Example, the string 111011 will be converted to a string 11111.

Note. All states of machines T1, T2 and T3, except for the start and final states are different.

5.2*. How many tape cells will be used (the used cells are those where machine head writes down any symbol at least once) by Turing machine T if number A consists of n units, and number B of m units?

6. Propositional logic and predicates. Work with manipulator "Tarski world". *Using this manipulator one can verify statements for figures standing on a rectangular chequered board; these statements use unary predicates: "is red", "is blue"; binary predicates "is side by side with", "is on the left from" (in the column located to the left), "is above" (in a row located above).*

6.1 Write down logical expression that describes only those configurations in which all the figures located on the same row, with red figures in the interior and blue figures on the bounds.

Answer. $\forall x \forall y (\neg(x$ is above $y))$ AND
$\forall x ((x$ is red$) \Rightarrow \exists y(x$ is side by side with $y)$ AND $(y$ is on the left from $x)$ AND
$\exists y(x$ is side by side with $y)$ AND $(x$ is on the left from $y))$ AND
$\forall x ((x$ is blue$) \Rightarrow (\forall y((x$ is side by side with $y) \Rightarrow$
$(y$ is on the left from $x)$ AND $(y$ is red$))$ OR
$\forall y((x$ is side by side with $y) \Rightarrow (x$ is on the left from $y)$ AND $(y$ is red$))))$

6.2*. Try to shorten number of quantifiers in the resulting logical expression and explain transformations.

4   Solution for the Task "Graphs"

To solve a task about graphs one should notice and formulate different graph properties. So, for example, if the degree of each vertex is k, then by multiplying the number of vertices V by k we get the number that is twice greater than the number of edges E: $kV = 2E$. This formula demonstrates that there exist no regular graphs with an odd number of vertices and an odd vertices degree.

The other important pattern is the Euler formula, it holds for planar graphs. If we draw a square shaped graphs, we will have 4 vertices, 4 edges and two parts in which this graph divides a plane. We will call this parts faces similarly to the polyhedrons, that may be represented on plane with these graphs: $V = 4$, $E = 4$, $F = 2$.

It is obvious that these numbers are dependent. How? If we draw a diagonal, the number of vertices will not change, and the number of edges and faces will increase by 1. If otherwise we add a vertex in the center of the square and connect it with one of its vertices by the edge, then the number of edges will not change, and a number of vertices and edges will increase by 1. So, one can notice that the sum $V + F$ and E increase similarly when new vertices and edges are added. Thus the expression $V - E + F = 2$ is constant. Our example implies that this constant is 2. The Euler equation: $V - E + F = 2$.

Using these consideration one can answer the task's questions: Let us substitute the first equation into the Euler's equation, we will have $2V + 2F - kV = 4 \Rightarrow 2F = 4 + (k - 2)V$.

Note that every face is surrounded with at least 3 edges, thus $3F \leq 2V$, because we may sum up a number of edges for each face and we will count every edge twice.

Let us return to the inferred equation and use the new inequality: $12 + 3(k - 2)V \leq 4E \Rightarrow 12 + 3(k - 2)V \leq 2kV \Rightarrow 12 \leq (2k - 3k + 6)V \Rightarrow 12 \leq (6 - k)V$.

This inequality can not hold for $k \geq 6$, thus $f(n) < 6$. This estimation is enough to answer the second question.

So, the maximal degree of a vertex in a connected regular planar graph is always less than 6. It is obvious that regular planar graph of degree 2 may be build for any number of vertices n.

It is not hard to design an algorithm to build regular planar graphs of degree three with an even number of vertices. (Remember, that it is impossible to have an odd number of vertices in this case). One should draw a regular n-gon. Then connect two opposite vertices with an outside edge. All other vertexes fall into pairs of symmetrical with respect to the axis defined by the first two vertexes. They should also be connected by edges.

It is also possible to invent a way to build regular planar graphs of the degree 4. One can combine two regular n-gons to obtain a regular planar graph with $2n$ vertices, this demonstrates that it is possible to build an even vertices regular planar graph starting from 6 vertices. But now we will show a way to build a graph for any n starting from 8. We begin with a left graph on the Fig 1, it has 8 vertices. Then each time we select a 4-sided face, there will always be one. We add a vertex inside and connect it with four vertices. Then we remove two initial edges of the face, see Fig 1.

It is impossible to have 5 vertices, because this will lead to a complete graph of degree 5 that is not planar. (One may use the Euler equation to show that:

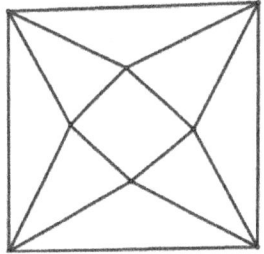 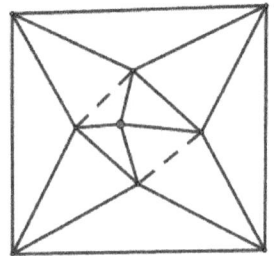 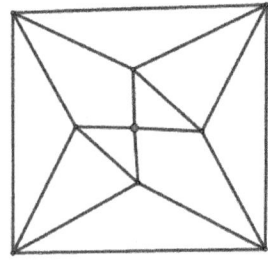

Fig. 1. Regular planar graphs with degree 4

$V - E + F = 2$, $5 - 10 + F = 2$, thus $F = 7$. Remember, that $3F \leq 2E$, but $21 > 20$).

It is also impossible to have 7 vertices. Let us show that. There should be $4 * 7/2 = 14$ edges in a graph. Thus, the number of faces is $F = 2 - V + E = 2 - 7 + 14 = 9$. So, each face has about $2 * 14/9 \approx 3,11$ edges. That means, we have all faces with 3 edges and one face with 4 edges. Let us start with the face with 4 edges, consider it is an outer face, then we may do a small search for variants that demonstrates it is impossible to build a graph (Fig 2).

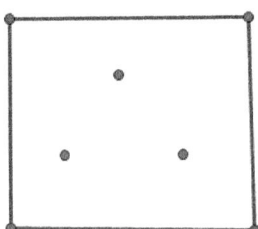 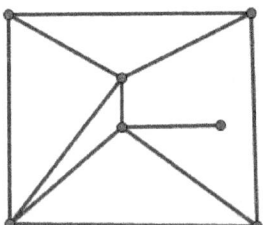 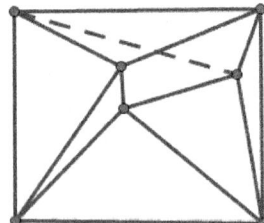

Fig. 2. Regular planar graphs with degree 4 and 7 vertices

Now we came to regular planar graphs of the degree 5. To demonstrate that such graphs exist, consider the graph of the icosahedron (Fig. 3). It has 12 vertices.

It is possible to build such graphs for other number of vertices by modifying the graph for an icosahedron. We do not know for what n such regular planar graphs exist, and we will stop here, because this is already more than enough for the problem. Let us only note that there is no such graph for 7 vertices, because 5 and 7 are odd.

Finally, we have that $f(n) < 6$ for all n, $f(n) >= 4$ for $n >= 8$. $f(1) = 0$, $f(2) = 1$, $f(3) = 2$, $f(4) = 3$, $f(5) = 3$, $f(6) = 4$, $f(7) = 2$.

5 Technology of DM&TI Tasks Design

The technology of tasks design includes several successive stages.

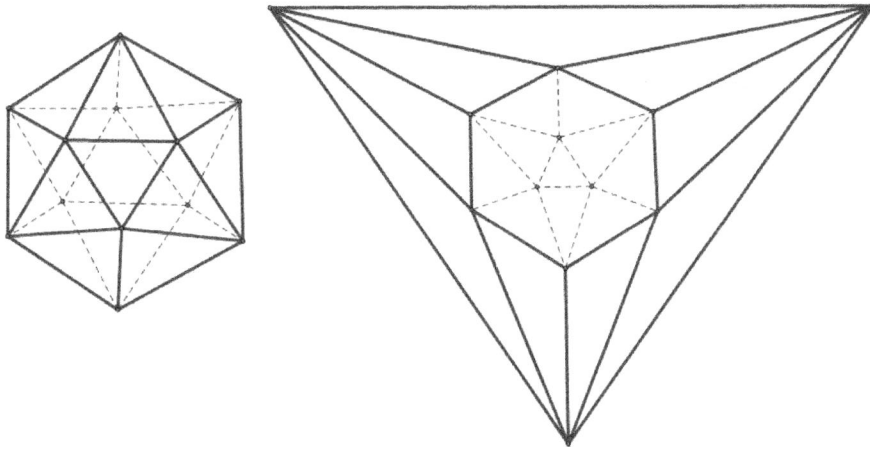

Fig. 3. Icosahedron's graph

Step 1. Select topic in computer science, which is considered as one of basic for theoretical computer science or discrete mathematics. At the same time, the topic should be accessible to be studied at the school level.

This does not take into consideration volume in which the theme was represented in the school. The knowledges will be justified by a format of Olympiad which allows to enter the student into subject domain gradually: for the first time new ideas were introduced on preparatory tour, next they repeated on extramural tour and finally on general tour, revealing the winner. All tours use the same computer manipulators, and Olympiad tasks base on concepts simulated by these manipulators.

Step 2. At this stage manipulators to support the participants activity must be designed and embeded to Olympiad site. The role of these manipulators is double. Firstly, they support student in theprocess of getting acquainted with the tasks statements and help students to understand an essence of the challenges. Secondly, the presence of the manipulator allows to explain the formal definitions of new concepts using the metaphor of the manipulator: it shortens the tasks statements.

Manipulator has several different functions:

– they are used as graphic illustrations of the tasks statements;
– they play the role of laboratory for experiments with math objects;
– they are used as interface to input solutions.

If the answer in the problem is the algorithm or expression that describes a class of objects, then manipulator includes a tools to verify students conjectures on the sets of examples as the participants have the opportunity to create new examples.

Examples

1. Correctness of constructed Turing machine was checked on examples; in addition to prepared examples, participants can enter any set of characters on the machine tape and step by step check machine work;
2. Correctness of the constructed regular expressions is checked on two sets of examples: first set contain correct strings, second - uncorrect ones. These sets of examples broaden by the participant. When participant checkes new regular expression, examples satisfying this expression will be highlighted.
3. The correctness of statements in the module "Tarski worlds" is checked on a set of configurations built by participants inside a graphics area of the manipulator.

Step 3. At this stage constructive and theoretical parts of every task should be formulated. The constructive part of the task is related to experimental work with the manipulator. Methodological role of this part is to prepare the student for more complex theoretical part of problem and to give an opportunity to collect experimental material for further generalizations.

The constructive tasks are the main type of tasks for the extramural round of the Olympiad. But for final round of the Olympiad theoretical problems become main type. The solutions of theoretical tasks participants gives in free form without using an automatic checking. The answer in theoretical tasks can be represented as in electronic or traditional printed form.

Step 4. At this stage must be developed and implemented various forms of feedback. On the first Olympiad tour (preparation tour) feedback can be done in the form of a hint when student input incorrect or incomplete answer. During the extramural tour going in real time a checking of partial solutions on some parameters can play role of feedback, which helps participants to complement or clarify solution.

Step 5. In this stage must be prepared all tutorial materials to demonstrate solutions for participants after the end of previous round of the Olympiad and serve as a means of preparing for the next round.

6   Technology of Manipulators Design

As mentioned above, the manipulator is essential part of methodological support of the Olympiad.

Manipulators:

- allow to perform experiments in the process of task solving;
- provide an environment for constructive tasks which help participants to quickly enter in the tasks topics;
- allow to formulate tasks statements by more clear language, based on the manipulator environment without formal mathematical definitions;
- help to formulate hypotheses for answers to theoretical questions by generalization of experiments and solutions of constructive tasks;- provide feedback in experiment progress;
- can be used as interface for inputting of solutions.

According to these aspects manipulator must provide the following opportunities:

− play a role of visualizer and have an intuitive interface;
− give possibility to construct complete or partial solutions;
− verify solutions on the sets of examples;
− give feedback to correct wrong or incomplete solutions;
− save solutions being made during round of Olympiad.

Let's now illustrate the manipulator device for tasks on logic circuits (Fig. 4).

Manipulator allows to construct any logic circuit from any number of given elements with given number of inputs and outputs. In the proposed manipulator to construct circuit we need to pull input or output of element to the desired contact. The appearance of a red dot indicates the connection has occurred. To remove the item, we drag it through the lower boundary of work area.

Manipulator allows the participant to verify the circuit at all possible input sets. For this purpose we use buttons at the top of the work area to get next binary set or go back. The signal flow through the wires is vizualized by color of conductors.

Constructed circuit can be saved as a task solution.

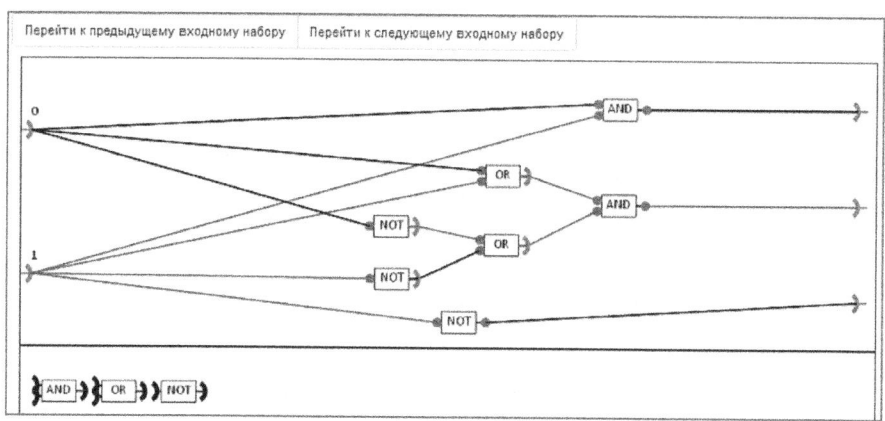

Fig. 4. The manipulator device for tasks on logic circuits

7   The DM&TI-2015 Results

In extramural round of the Olympiad participated about 350 students, about half of them took the opportunity to participate in preparation tour. Extramural tour consisted only of constructive problems, three hours for many of them appear to be not enough to find solutions for all six problems, so the internal round took only 47 people. Participants of internal round are required to solve 6 constructive problems with 3 points for each, and 6 theoretical problems with 6 points for each (each theoretical task was based on corresponding constructive task). The following rules were used for rewarding:

- participants with at least 24 points scores ought to be awarded a diploma of 1 degree,
- participants with scores from 18 to 23 points ought to be awarded a diploma of 2 degree,
- participants with scores from 12 to 17 points ought to be awarded a diploma of 3 degree. Results of DM&TI-2015: diplomas of first degree - 0, diplomas of second degree - 2, diplomas of third degree - 7.

8   Conclusions

Analysis of all DM&TI-2015 rounds of the Olympiad result in following conclusions:

1. Presence of manipulators in theoretical tasks allows to engage in solving tasks those students, who still not possess the ability to work with problems in formal-theoretical formulation. Constructive problems are solved by ten times more participants than the theoretical ones.
2. Using of the same manipulators in different rounds of the Olympiad allow participants to enter in a subject area gradually, prepare to solve theoretical problems through experiments and solution of constructive tasks.
3. Constructive tasks on manipulators with logical circuits and logical statements are used by teachers in the classroom and extra curricular activities and assessed them as a help in the presentation of questions in theoretical computer science. As to tasks in discrete mathematics, teachers give ambiguous responses. For example graph conception some teachers are not considered as a topic of computer science; some teachers suggested to find a more practical tasks in discrete math.

References

1. Hromkovi, J.: Theoretische Informatik. Formale Sprachen, Berechenbarkeit, Komplexittstheorie, Algorithmik, Kryptographie, 5th edn., 349 p. Vieweg+Teubner (2014) ISBN: 978-3-6580-6432-7
2. FES: Secondary (full) general education. FGOS: Srednee (polnoe) obshhee obrazovanie, http://standart.edu.ru/catalog.aspx?CatalogId=4099
3. Pozdniakov, S., Gaisina, S.: New trend in Russian informatics curricula: integration of math and informatics. In: Local Proceedings of the 7th International Conference on Informatics in Schools: Situation, Evolution and Perspectives, pp. 91–100, http://www.issep2014.org/wp-content/uploads/2014/09/issepr _2014_proceedings_book.pdf

4. Pozdnjakov, S.N., Fomin, S.V.: Pervaja Vsesojuznaja matematicheskaja olimpiada uchashhihsja srednih proftehuchilishh (zadachi jeksperimental'nogo tura). zh. Matematika v shkole, N2 (1986)

5. Bashmakov, M.I., Pozdnjakov, S.N.: Matematicheskie olimpiady v srednih proftehuchilishhah. In: Prosveshhenie, M. (ed.) Biblioteka Uchitelja Matematiki. Matemati-cheskie olimpiady (1988)

6. Carteli, A., Dagiene, V., Futschek, G.: Bebras Contest and Digital Competence Assessment: Analysis of Frameworks. International Journal of Digital Literacy and Digital Competence 1(1), 24–39 (2010) ISSN 1947–349-4

7. Pozdniakov, S., Posov, I., Pukhov, A., Tsvetkova, I.: Science Popularization by Organizing Training Activities Within the Electronic Game Laboratories. International Journal of Digital Literacy and Digital Competence (IJDLDC) 3(2), 17–31 (2012)

8. Pozdnyakov, S., Posov, I., Akimushkin, V., Maytarattanakon, A.: The bridge from science to school. In: 10th IFIP World Conference on Computers in Education, WCCE 2013, Torun, July 2-5 (2013)

9. Tedre, M., Moisseinen, N.: Experiments in Computing: A Survey. The Scientific World Journal 2014, Article ID 549398 (2014)

10. Naps, T.L., Rößling, G., Almstrum, V.L., Dann, W., Fleischer, R., Hundhausen, C.D., Korhonen, A.: Lauri the role of visualization and engagement in computer science education. ACM Sigcse Bulletin 35(2), 131–152 (2003)

11. Ilya, P., Sergei, P.: Implementation of virtual laboratories for a scientific distance game-competition for Schoolchildren. In: The 2013 International Conference on Advanced ICT (Information and Communication Technology) for Education (ICAICTE 2013), Hainan, China, September 20-22 (2013)

12. Contest Management System, https://github.com/cms-dev/cms

13. Akimushkin, V.A., Majtarattanakon, A., Pozdniakov, S.: Tehnologii avtomatizacii raboty s issledovatel'skimi zadachami na primere zadachi Chasy-kalendar. Izvestija SPbGJeTU LJeTI 4, 34–41 (2014)

14. Akimushkin, V., Korepina, I., Puhov, A.: Distance school in discrete mathematics: learning algorithms on graphs. In: The 12th International Congress on Mathematical Education (ICME 2012), July 8-15. COEX, Seoul (2012)

CS Unplugged: Experiences and Extensions

Irena Demšar[1] and Janez Demšar[2]

[1] Alojzij Šuštar Primary School, St. Stanislav Institution, Ljubljana
[2] University of Ljubljana, Slovenia

Abstract. CS Unplugged is a set of activities for teaching CS concepts without using computers. We translated it to Slovenian and used it in different contexts, from the classroom and afterschool activity to summer school to professional development courses. In the paper, we summarize our adaptations, extensions and experiences.

1 Introduction

CS Unplugged (`csunplugged.org`) is a set of activities developed in mid-nineties by Tim Bell, Ian Witten, and others. It builds on the premise that computer science is not about computers (just as astronomy is not about telescopes) and does not need to be taught with computers, which are often just distracting. As the lack of skilled engineers, especially in the western hemisphere, is becoming more and more acute, the school curricula are shifting from teaching basic literacy and use of technology back to teaching the CS concepts, which leads to a deeper understanding of technology and, consequently, prepares pupils to eventually become its creators and not only consumers. With this paradigm shift, CS Unplugged is gaining traction in many countries around the globe.

1.1 The Slovenian Translation

We translated the activities to Slovenian language (`vidra.si`). One of the authors (ID) is a primary school teacher with 2^4 years of teaching experience, and the other (JD) teaches at the Faculty of Computer Science.

Our translation is augmented by several new activities, and some original activities are extended or renewed (*e.g.*, finding a phone book in 2010s may be difficult). Some changes were inspired by the material from MATHmaniaCS (`www.mathmaniacs.org`), cs4fn (`www.cs4fn.org`), Chris Bishop's Christmas lectures (`www.richannel.org/christmas-lectures/2008/2008-chris-bishop`) and similar. Some changes were influenced by the equipment and materials that are typically found in (or missing from) Slovenian schools. Finally, we expanded the explanation of the CS background in hope to also attract teachers without much (or any) knowledge of computer science. We try to approach them – and abate their fears – by presenting CS as a somewhat formalized everyday reasoning.

We added more kinesthetic approaches and prepared alternative setups that allow more activities to be performed outside. We emphasized the role playing

A. Brodnik and J. Vahrenhold (Eds.): ISSEP 2015, LNCS 9378, pp. 106–117, 2015.
DOI: 10.1007/978-3-319-25396-1_10

aspect. For instance, in the original Deadlock activity the children pass fruits to each other, while in our setup children themselves walk around, by which they get more literally deadlocked. Another emphasized feature are puzzles, surprises and magic tricks; these tickle the children's brains. We sometimes stray from CS towards pure mathematics or general problem solving.

We also shifted several activities towards gamification. We introduced competitions as a form of motivation. These also promote team work and sometimes require the group to organize. An example of such activity is the decoding of the text written in binary. An even more interesting game is the already mentioned adaptation of the deadlock activity, which allows for observing (and hopefully improving) the social dynamic of groups.

1.2   Setup

We are gaining experience and getting new ideas by conducting these activities in different contexts. We first started with a few activities in the classroom during mathematics in the fourth grade. These included binary numbers and sorting algorithms as well as some newly developed activities. In 2012/13, we led an afterschool activity (at ID's school) that covered most of the CS Unplugged material. We are also in contact with another teacher who organized a similar activity for pupils in the second grade (7-8 years). She adapted some activities and did not cover the later, more algorithmic parts, but otherwise reports that the activities were very popular among children.

In the summers of 2013 and 2014, the authors of the paper and a group of around ten volunteers from the Faculty of Education and the Faculty of Computer Science, University of Ljubljana, organized summer schools for children from K4 to K6. Summer schools were free of charge (except for the small fee to cover the meals). In the first year, the school was attended by 42 children and in the second year we limited it to 30.

We also use parts of this material for other promotional events, like the Days of computer science at the Technical Museum Bistra, promotional activities in high schools, events at science festivals and similar.

Finally, we organize professional development courses for teachers, in which the material is presented in the allotted time frame of 24 hours. We have also prepared smaller-scale presentations and local conference workshops. The presentation is "hands-on": teachers learn the activities in the role of the children, after which we explain the computer science background and give some recommendations and warnings regarding the execution of the activity.

2   Experiences and Adaptations

We shall review some activities performed in the contexts listed above. The review is not systematic as we bias it more towards interesting new ideas and anecdotes. We assume the reader's familiarity with the CS Unplugged activities.

2.1   Binary Number Representation

Our first session begins with attracting the attention by asking the group about the largest number that they can show with their hands – and we act surprised when they answer 10. We promise to show them how to count to much more. We then proceed with the cards with dots from the CS Unplugged activities.

We always continue the activity so that the attendees (children or adults) make squats, either holding the cards or, later on, without them. We ask them *which number between 0 and 31 cannot be represented in such a way;* the question *whether it is possible to represent all numbers the answer* would be too suggestive, while the *which number* formulation tricks the children as well as adults into thinking.[1] After some guessing – and trying these numbers out, always discovering that we can represent any number they suggest – they infer that probably all numbers can be represented like this.

The obvious way to prove this is to count from 0 to 31 by making squats. They realize that the person representing the lowest bit will get muscle fever while the highest will not break a sweat. This can lead to discussion about odd and even numbers. We discover an update rule (the lowest bit goes up and down, while everybody else changes his position when his left neighbour goes down).

We continue this by pondering whether this also holds for higher numbers and suggest adding five more bits to the existing five. The children are amused to realize that the lowest bit would then need to make 512 squats instead of just 16. In the later activities, we often refer to this finding, as it is strongly related to the exponential time complexity as well as, on the other side of the coin, the logarithmic complexity of bisection.

We then learn to count to 31 with fingers of one hand, for amusement and for later use. In the second session of the afterschool activity, one girl had a bandage on her second finger due to sport injury. We asked her to show number 17. She could not, which led to an interesting discussion about which number she can show and which not. Throughout our session in afterschool activity, the lessons often included a lot of mathematics, which was not the case in the summer school. The reason may be that the former took place in the school and with a smaller group of calmer children, while the latter took place during holidays with a larger group of lively (and louder) children.

Making squats is also an excellent icebreaker when presenting the material to adults.

In professional development courses, we emphasize that the purpose of the activity is not to learn and practice the conversion between the binary and the decimal representation, but to understand the essence of the binary system and its use. The take-home message should be that the binary system is practical and used in computers since it allows us to represent any number with a sequence of two-state objects or events and that each additional bit doubles the range (with important consequences for later discussion of time complexity). We find parallels of these properties in the more familiar decimal system.

---

[1] Another, even more conducive question (although it represents only the other side of the same coin) is *which number can be represented in two different ways.*

2.2   Text Encoding and Decoding

To put storing the numbers in binary format to use, we next show how to encode or transmit one-word messages by a sequence of two-state objects or events. When we first tried it during the afterschool activity, the children wrote down a four-letter word to one of the authors, who "transmitted" it to the other author by singing a jazzy song similar to those in the "Modems Unplugged" activity. We later expanded this and the puzzle about the trapped Tom into a larger activity, in which one child in a group is given a four-letter word and has to transmit it to other members of the group using the means we provide, such as a set of black and white pieces of paper hung from a rope by clothespins, clapping and thumbing, banging a smaller and a larger pot with a wooden spoon and, in general, making noise in various (binary) ways.

We have encoded the first two paragraphs of Professor Branestawn stories into 0s and 1s and put it onto 180 enumerated strips of paper, one word on each. We split the children into four groups of ten children, gave each one set of papers (thrown on the desk to shuffle them) and let them compete which group will be the first to decode the entire text. We underestimated the effort needed; the task required around one hour. Fortunately, we also underestimated the children's perseverance and the effect of gamification. When parents came to pick them up, the children told them to wait until they are done decoding.

The task is a good example of how the binary system *should not* be taught, *i.e.* by training the mechanical conversion back and forth. Yet it was instructive for another reason. One group had two very bright children who soon learned to read most of the letters by just looking at the binary code (e.g., they learned that 01010 was the letter I), yet the group finished the last due to poor organization. Other groups pipelined the process and, in particular, organized the sorting. One group, following a tip by one of the instructors, divide-and-conquered the task of sorting the words by splitting them into three groups with numbers 1–60, 61–120 and 121–180, which they sorted by dedicating three team members to this bottleneck of the decoding process. Another group wrote the numbers from 1 to 180 onto a few sheets of paper and instead of sorting the paper strips they just wrote the decoded words beside the corresponding numbers – essentially inventing a random-access data structure.

While the task missed the intended point, it has hit another one: the groups applied some concepts from computer science, like pipelines, divide and conquer, data structures with random write access to solving a real-world problem – which is the essential purpose of teaching algorithmic thinking to general population. They also learned that organization – the architecture of the system, in CS terms – might matter more than speed of its parts.

We ourselves learned about the unbelievable persistence of children and the power of gamification.

2.3   Error Correction

We usually introduce error correction by the magic trick suggested in the CS Unplugged. In the summer school though, the activity was introduced by a brilliant show by two world-famous telepaths. Some children bought it, while others considered it a good trick. Children also challenged one of the authors (JD) whether he too possesses any telepathic powers. This coincided with their idea about turning two cards instead of just a single one, which makes the error correction ambiguous. He confessed that he is not much of a telepath and was not sure about how accurately he can do it...It turned out that the telepath, to whom JD sent the information "did not hear well and could not decide between two possible changes".

The activity continues by teaching the children how to perform the trick and practicing it with their schoolmates. In the afterschool activity, some children posed interesting questions. The bottom right card corresponds to the parity of the control row and, at the same time, the control column, so they were interested if there can be a conflict between these two. One girl found an excellent constructive proof that this cannot happen: she started with a matrix of 0 and then said that whichever configuration we may want to set up, we can turn one card at a time and, consequently, turn the corresponding cards in the control row and column, and also the bottom right card. As no conflict can arise within this operation, there can also be no conflict when we reach the desired setup.

Another question was whether we can control oddity instead of parity. We did not know the answer and gave it in the next session. Creative children can easily ask questions that a doctor of computer science has not considered before.

Unlike the original CS Unplugged activity, we do not continue with ISBN codes, but with 13-digit EAN, which are now more common and also easier to compute. We gave the children various item like boxes and wrappings instructed them to compute the EAN checksums (across all 13-digits, including the control) without telling them what the result should look like. One of instructors pretended to be specially skilled in fast calculation; he was able to check their results in a bliss, rejecting the wrong ones (after trying again, children found he was always right). He kept writing the correct ones on the blackboard, until the children spotted the pattern, that is, that all correctly computed numbers are divisible by ten. Note again the element of magic and gamification.

We rounded up the activity by explaining that, similarly to the cards in the parity trick where the last card in a row or column does not contain any information, bar code readers also use the last digit to verify the correctness of the code.

If the time permits and the children are interested in mathematics, we show how to verify whether the sum of two large numbers is correct by adding the digits of the summands and the digits of the sum. Children often respond by verifying whether the trick also works on "really large" (say 20-digit) numbers (they apparently perceive a verification with large numbers to be more convincing). We explain the background as deeply as they want, which typically goes up to showing why the sum of digits of a number gives the remainder of its division by 9. This "trick" belongs more to math than to CS, yet it is related to the main activity and the children find it useful (one child complained that the math teacher should have told them about this).

2.4   Text Compression

We have shown the original version of the text compression activity to individual children, who did like it and enjoyed constructing other sentences that compress well. We have however found it difficult to demonstrate that the compressed text indeed takes less space than the original; the compressed string BAN(2, 3) looks larger than the original, BANANA.

Tim Bell told us that another author of CS Unplugged, Mike Fellows, used to draw Huffman trees on the ground and let the children "decompress" messages. We thus prepared a few dozens of roughly 20-character sentences and encoded them using a Huffman tree, a Morse tree and a binary tree of depth 5 with letters arranged from left to right. Each sentence is encoded by a sequence of letters L and D (short for left and right in Slovenian). We draw one or two trees of each kind on the ground and let the children decode messages by walking on them.

It is difficult to walk around while not losing track of the sequence at the same time. Children themselves solved this by forming pairs, in which one child is giving instructions (left and right) and the other is walking and shouting out the letters when (s)he reaches a leaf. The obvious potential for quarreling made the activity even more fun.

The limiting factor here is the physical size of the trees; smaller trees cannot accommodate a large number of children, while larger trees may soon occupy too much space. On several occasions we performed this activity indoors. Instead of drawing the tree we used painter's duct tape, which is cheap and can be removed without leaving any marks. In the future, we are considering marking the tree branches with ropes and tent pegs, which will allow for larger trees, possibly on a hill, which would allow us to turn decoding into an exhausting running competition. This may not directly result in a better understanding of text compression at all, but it will add some sweat, mud and fun.

This activity and the error correction follow the text encoding; we justify them by observing that communication, for instance by singing the high- and low-pitch tones, is error prone and takes too much time, hence the need for error correction and compression.

2.5   Programming Languages

We put the activity about programming languages before other algorithmic activities in order to emphasize the importance of precise instructions. When we later play with, for instance, sorting algorithms, we ask the children to exactly describe the algorithm they discovered (or else we follow the instructions they provide and misunderstand them wherever possible[2]).

Although the activity involves just drawing simple pictures, it was always popular with children of different ages as well as teachers attending professional development courses.

---

[2] See http://youtu.be/QZL2esc8Hso from the Educational Robots for Absolute Beginners CS4HS MOOC for an example.

At the summer schools, we formed groups of four children in which one of them was giving instructions to others. On other occasions, we kept a single group and let a single person give instructions to all participants. The children who follow instructions often grow impatient and believe that the instructions are unnecessarily confusing, not realising the difficulty of the job. We came up with the following twist: instead of showing the group the original picture after they finish drawing, we invite the loudest heckler to do a better job. Their typical first reaction is: "Oh, that! This is simple." It often happened, though, that the picture (s)he has initially drawn was also consistent with his (supposedly better) instructions. We have sometimes re-drawn the same picture for four or five times before showing the actual picture to the group.

We conclude with a discussion about programming languages. Especially when working with older children and adults, we draw their attention to the vocabulary: for some pictures they use terminology from geometry or general knowledge (like "David's star"). We note that different programming languages have different syntax reflecting their different intended uses.

We developed another activity related to programming languages. We prepared cards for simple set of word transformations (removing and adding letters, reversing words and similar).[3] We show how to turn an apple into a sausage (JABOLKO and KLOBASA in Slovenian). Then we distribute cards and assign the participants a few transformation tasks.

We then introduce "computing machines" in the form of trees that represent arithmetic expressions.[4] We arrange the children in a tree. Leaves have a role of an input, random input or a constant, and internal nodes perform operations (addition, subtraction, multiplication) up until the root announces the result.

After some training we let the children construct a machine for computing the perimeter of a rectangle. When trying this at school, children constructed a machine that computed the perimeter as $2 \times a + 2 + b$, with the children "a" and "b" having the role of measuring the longer and the shorter side. The machine is incorrect ($+$ instead of $\times$), although the children wrote the correct formula on the whiteboard. Instead of telling them, we used the "malfunctioning" device to compute the perimeter of an A4-sheet of paper (3 by 2 dm). The result was correct since $2 \times 3 + 2 + 2 = 2 \times 3 + 2 \times 2$. We then gave them one third of the paper (2 by 1) to discover that the program, which works once, does not necessarily work for all inputs. Initially they assumed that some participant made a wrong calculation, then they checked the tree step by step until discovering the error. Their accidental mistake led to an unplanned lesson about debugging, as well.

Months later, these children encountered tree-like representation of arithmetic expressions during a regular mathematics class. Even those who are otherwise underperforming grasped the concept without needing any special explanation.

---

[3] Inspired in part by the task Text machine from the Beaver competition 2012; see *e.g.* www.beaver-comp.org.uk/uploads/2/1/8/6/21861082/bcccontest2012.pdf.

[4] Inspired by the activity All very logical (Chris Bishop, Christmas Lectures 2008), www.rigb.org/christmaslectures08/html/activities/all-very-logical.pdf.

2.6 Bisection

We used the Battleship activity on a few occasions, but later replaced it with a simpler and shorter one, which introduces bisection. Fifteen children are holding hidden sheets of paper with numbers, and another child gets five candies. His job is to find some given number. The children holding the numbers reveal the number only if they are paid a candy. We try this exercise with numbers in random order or with ordered ones. Children soon discover that their chances of keeping any candy at all in the former case are slim, while in the case of sorted numbers they easily discover the bisection algorithm.

We then relate bisection to the binary number representation. We play "guess the number" so that one child thinks of a number and the instructor's job is to guess it. Instead of the usual "is your number ≥ 8", we tell them to show the number with their fingers, while hiding the hand under the desk. We then ask about the state of each finger. This makes it clear that the number of questions needed to guess any number equals the number of fingers needed to show it.

In discussion, we explain why asking about the thumb is the same as asking whether the number is higher or equal to 16 (and so forth). Since we know – from the activity about the binary numbers – that each additional digit allows for doubling the range of numbers, the same holds for bisection: an extra candy doubles the range of numbers, which, told from the other side, means that doubling the number of children in the motivation example would require just a single extra candy. To further emphasize the power of bisection, we note that if they would mark a single atom in the visible universe, we would be able to "guess the atom" in roughly 250 guesses (assuming that the children already know about atoms).

While the notion of logarithmic time complexity can remain on the intuitive level, the discussion during the afterschool activity went so deep that we gave in and introduced the word "logarithm" to our vocabulary. We defined it as the number of times we can divide a number by two before we get to 1 or less.

2.7 Sorting Algorithms

In the activity related to sorting algorithms we replaced film canisters by plastic yogurt bottles, which are abundant in schools. Since they can hold more water, the differences in weights are larger, and we can use imprecise scales made of a plank with a nail glued at its middle.

When we tell the children to sort a set of bottles, they will almost always discover selection sort (and, in rare cases, insertion sort).[5] This is fortunate since the number of comparisons, if performed correctly, is always equal. We find the reason for this, derive a general rule and compute the number of comparisons for different number of bottles to discover its quadratic behaviour. With older children (K7 and above), we introduce a general formula $n(n-1)/2$ and reason

---

[5] While bubble sort is often discovered by programming beginners who try to invent a sorting algorithm, its limitations are not natural when the objects are not stored in cells (arrays). We never saw children discover it when sorting bottles on a desk.

about why the only important part is n^2. We thus introduce the intuition behind the asymptotic time complexity and the big-O notation.

After letting the children to (unsuccessfully) try to invent a faster algorithm, we show them the first step of quicksort. We propose to split the bottles into the lighter and the heavier than a randomly chosen bottle. (For a greater effect, we cheat by remembering which bottle was in the center of the sorted sequence before randomly shuffling them.) We put the reference bottle in between the groups and note that the bottles are not ordered yet. At this point, the children *always* suggested repeating the same procedure on both sides. The children are therefore not afraid of recursion (in contrast to CS students). We let them practice the algorithm and count the number of comparisons.

One girl asked us about the time complexity of quicksort. We told her that the number of comparisons depends upon chance and computing the average is difficult, but promised to explain the worst case later on. After a few seconds she said: "Oh, I know, it's the same as before." (We did not measure the time an average CS student would need to realize this on his or her own.)

We occasionally also try bubble sort. We use clothespins to attach numbers to children's shirts and we put them in a line. They then pass a stick from left to right, and the two children holding a stick swap their places if the left one has a lower number than the right.

Trying to show merge sort was a great failure. Showing a single step of merging two sorted runs into a single one works. Merge-sorting 16 children using a parallel algorithm – the story is that we have 16 lanes merging into one, with 15 policemen directing the traffic – does not work due to traffic jams occurring in the later merges. Unfortunately, we tried this for the first (and last time) with not 16 but 40 children on a parking lot in a simmering hot summer afternoon.

2.8   Sorting Networks

We gamified the sorting networks activity. We split the children into groups of six and give each child a number. Teams are marked with different colors. Instead of walking across the network drawn on the ground, we use the trees in the park as nodes. We marked them with numbers in no particular order. At each tree we put the instructions to wait for the team mate (somebody with the same color mark); after they both arrive, they compare their numbers and this determines the tree to which each of them should run. The final nodes are on the parking lots, one color-coded lot for each group. After the signal, the children start running between the trees. When all member of the group arrive, we check whether they are sorted correctly (which looks like magic); if they are not, the whole groups has to start again.

The activity in this form does not teach them about sorting networks: it serves as a fun game with a lot of running, but also makes them curious about how this actually works. To satisfy the curiosity, we introduce them to sorting networks which we draw on the ground in advance, let them construct their own networks and so forth.

2.9   Deadlock

Deadlock is one of the favourite activities. We do not use the activity as proposed in the CS Unplugged but its equivalent from MATHmaniaCS, in which we have a graph with 9 nodes, 7 of which are colored with unique colors and 2 are empty.[6] We form groups with 7 children. Every child in a group is marked by one of the nodes' colors. We put the children onto nodes in random configuration. They are allowed to move between the connected nodes, but only if the destination node is free. The goal is achieved when everybody reaches the node of his (her) color.

In the competition, they soon realize that they can only win as a team and an individual member may need to sacrifice his achieved (sub)goal to let the others reach theirs. After a few rounds, we spice it up by prohibiting any form of communication. Next we constrain them by introducing a "token": only the person holding the token may move. The token can be passed arbitrarily between the players, and they are allowed to communicate. Finally, they must solve the tasks with the token and without communication. The token is already proposed in MATHmaniaCS, and the no-communication rule is our idea that forces the children to consider the needs of the others and give way to them without being asked. Working in silence forces the group to come up with a common plan and cooperate without being able to discuss it.

We once had a problematic boy who happened to have the color of the most central node. After getting there, he refused to give way to others and the group was unable to solve the task at all.

We tried the game during a gym class, first with girls and then with boys. The girls planned and solved the task with ease. The boys, who have not seen the approach of the girls, quarreled and moved randomly, until one of them took the role of the leader and directed the others – without success until realizing that he himself has to move from the central node to let the others pass.

Children like this game. Many teachers also found it interesting due to its cooperative problem solving aspects. We also tried this game with a group of children attending an (non-CS Unplugged) event for socially underprivileged children. The result showed they were far less collaborative than other groups, showing the effects of living in such social environment.

2.10   Algorithms on Graphs

We use the minimal spanning trees problem to stress the importance of giving and following a formal algorithm description. Furthermore, we show how graphs can provide a useful, cleaner abstraction of the data.

The aspect of a graph as an abstraction is even more evident in graph coloring algorithms. For the initial motivation, we give the children the task of scheduling afterschool activities. We find a solution without using any particular algorithm. We then introduce the problem of map coloring and abstract the maps into graphs. We return to the scheduling problem and show its solution using map

---

[6] mathmaniacs.org/lessons/16-deadlock/index.html

coloring. Finally, we show the relation between graph coloring and sudoku (we use a 4-by-4 color sudoku). Our aim is to demonstrate how the same algorithm can be used for solving multiple seemingly unrelated problems.

Before this activity, children already meet graphs for several times. Scheduling is the first problem in which the nodes do not correspond to some physical places but to more abstract concepts (afterschool activities). We did not feel that 10-12 year old children have any problems with this abstraction.

When discussing the dominating sets problem, we recapitulate the time complexities, from bisection with logarithmic complexity, random search with linear complexity, sorting with quadratic complexity to the NP-complete dominating sets. We do not explicitly list graph coloring as an NP-complete problem since the children find the task of graph coloring an easy one (while, of course, using simple heuristics and guessing to find the optimal solution for a simple graph). Dominating sets, on the other hand, are a clear example of a difficult task.

Graph algorithms conclude the arc from binary numbers and various encoding and compression tasks, to giving instructions, to algorithms, to time complexity. The remaining topics present two specific areas of computer science.

2.11 Cryptography

Cryptography is an interesting and popular subject. We expanded the related activities to several hours, in which we show different cyphers, including the Vigener cypher using a wheel that we designed for this purpose.[7]

We also use the Sharing secrets for short introductions to the essence of computer science. The original activity suggests income as the personal detail that we do not wish to reveal. This is does not work well most age groups. Instead, we tell them they want to use an elevator so they have to work out their total weight. Since we typically do this in presentations in highschools and for older audiences, we try to involve girls or women, to play on the stereotypical premise that they are not ready to reveal their weights. We spend plenty of time trying to figure out the solution, so we can, in the end, after discovering it, present the computer science as the art of finding solutions for seemingly unsolvable problems.

2.12 Artificial Intelligence

The CS Unplugged activities related to artificial intelligence are focused on the discussion about what does the term AI actually mean. The motivation for the discussion is an "intelligent piece of paper" that never loses a game of tic-tac-toe. We lead the children into saying that it is not the paper that is intelligent, but the person who wrote it. They admit that a computer that would learn such a strategy by itself would indeed be intelligent.

This naturally leads to an activity, which is not described at the CS Unplugged: the Sweet computer,[8] originally by Martin Gardner, but inspired by

---

[7] vidra.si/javna-gesla/Pripomoček za Cezarjevo in Vigenerjevo kodo.pdf
[8] www.cs4fn.org/teachers/activities/sweetcomputer/

the AI pioneer Donald Michie. The children "construct" the computer and then they play against it until the computer, which initially draws dumb moves at random, becomes unbeatable.

Since this computer learns to play, it is, by their own definition, intelligent. Even younger children become quite involved in discussions about why this still is not a "real intelligence" and argue that the one who "programmed" the computer to learn did the intelligent job. With adults, we conclude by the Searle's Chinese room argument and the Wittgestein's Beetle in the box.

3   Conclusion

CS Unplugged is an excellent inspiration for teaching computer science concepts. It can be used for children as young as 6, as well as for adults. Even the CS students who already absolved the typical courses in algorithms and data structures said that exposure to these activities and thinking about those concepts in real-life-like scenarios have deepened their understanding of the subject.

Our adaptation adds even more kinesthetic learning and gamification. Both are particularly important in the context of summer schools. Moreover, CS Unplugged is planned to be included in the elective subject "computer science" in the 5th grade of primary schools, starting in 2015/16. The adaptation is aimed particularly at teachers without a proper training in CS. While this may provide a starting point for them, teaching concepts is much harder than teaching facts and recipes like the conversion between binary and decimal number representation: it requires a broader and deeper knowledge of computer science, to allow the instructor to answer questions and to let her or him be led by the children's interests and questions, which is the only way to turn the lesson into an inspiring and captivating experience.

Three years ago, two very gifted 11-years old girls were not interested in computers and refused to join the CS afterschool activity. We tricked them by hijacking another afterschool activity and presented the CS Unplugged material there. They enjoyed it and later participated in the summer school. Next year, they attended activities based on Scratch and the summer school about AppInventor. This year we have, upon their request, taught them Python. They achieve excellent results at Bebras competition, they explain recursion to their younger mates, read CS-related popular books and want to use PGP in their e-mails. And they say they know what they wish to be when they grow up.

Acknowledgments. We would like to thank the students and children who participated in organization of described events and thus helped shaping the activities. Particular thanks to Teja Šavs who is the co-author of the new activities described in section 2.5.

Computing at School in Sweden – Experiences from Introducing Computer Science within Existing Subjects

Fredrik Heintz, Linda Mannila, Karin Nygårds, Peter Parnes, and Björn Regnell

Linköping U., Linköping U., Sjöstadsskolan Stockholm, Luleå U. of Tech., Lund U., Sweden
fredrik.heintz@liu.se, linda.mannila@abo.fi, karinnygards@gmail.com,
peter.parnes@ltu.se, bjorn.regnell@cs.lth.se

Abstract. Computing is no longer considered a subject area only relevant for a narrow group of professionals, but rather as a vital part of general education that should be available to all children and youth. Since making changes to national curricula takes time, people are trying to find other ways of introducing children and youth to computing. In Sweden, several current initiatives by researchers and teachers aim at finding ways of working with computing within the current curriculum. In this paper we present case studies based on a selection of these initiatives from four major regions in Sweden and based on these case studies we present our ideas for how to move forward on introducing computational thinking on a larger scale in Swedish education.

1 Introduction

Computing is no longer considered a subject area only relevant for a narrow group of professionals, but rather as a vital part of general education that should be available to all children and youth. The December 2014 issue of ACM Inroads featured a special section on early computing education. The articles highlighted several questions related to the when, what and how of introducing computing prior to university level [1].

In October 2014, the European Schoolnet published a report on the current status of computing at schools in 20 European countries [15]. According to the report, a majority of the countries are introducing computing (or programming) at primary and secondary level, either as a subject on its own (e.g. England), in specific IT courses (e.g. Belgium and Estonia) or as an interdisciplinary strand in other subjects (e.g. Italy and Finland). In several countries, this work is, however, still in a very early phase, and generally there is no clear consensus on what computing education at primary and secondary level should entail or how it should be introduced into the education system.

Since making changes to national curricula takes time, people are trying to find other ways of introducing children and youth to computing. In Sweden, several current initiatives by researchers and teachers aim at finding ways of working with computing within the current curriculum. While focus in several countries is put on programming, we believe this to be too narrow focus. If you only focus on programming and code, you risk missing out on general and useful skills such as dividing problems in smaller parts, solving problems in creative ways, finding patterns, thinking logically, designing algorithms, working in a structured manner, making generalisations and finding models.

A. Brodnik and J. Vahrenhold (Eds.): ISSEP 2015, LNCS 9378, pp. 118–130, 2015.
DOI: 10.1007/978-3-319-25396-1_11

Similar to many others (e.g. [2,17]) we use the term "computational thinking", rather than coding or programming, to refer to this set of skills and practices. Teachers already engage in several of these practices in their teaching, even without knowing it [7]. By making explicit what teachers already do, we can reduce the feelings of threat to teachers who may feel that they need to learn a lot of new things, many of which can be perceived as technically advanced.

The aim of this paper is to give an overview of some of these initiatives to give examples of how teachers can start working with computational thinking in their classroom, without any official policies requiring them to do so. We start by giving a brief overview of the Swedish school system and the current status of computing in K-12 education. Next we present a set of national and regional initiatives. The paper is concluded by some reflections on the current work as well as ideas for future directions.

2   Overview of Computing in Swedish Schools

The Swedish education system is divided into compulsory school (grades 1-9) followed by upper secondary school (grades 10-12), which offers 12 vocational programmes and 6 programmes preparing for further studies. The national curriculum for compulsory education does not include computing or programming. One of the programs of the general education strand is focused on Technology (*teknikprogrammet*), which also includes courses covering various aspects of computing. This means that only a limited number of students take such courses at upper secondary school. All students do, however, get to use technology, i.e. computers and applications, in their studies.

The situation in Sweden is hence similar to that of many other European countries, where digital competence has come to be seen as something of a synonym for basic digital literacy. Whereas an increasing number of countries are revising their curricula and policy documents to also include computing aspects, Sweden has a rather new curriculum, which most likely will not be updated within the near future.

In 2012, the Swedish government established a committee with the task of giving recommendations and guidelines for how Sweden can – and should – benefit from the digitalisation. In a report published in March 2014 [3], the committee emphasises the need for an additional focus on digital competences in national curricula. One concrete recommendation is for programming to be introduced as a cross-curricular element in already existing subjects.

3   Bebras

Since computing has not traditionally been part of general education, other, more informal, approaches have been introduced in order to cover the same ideas. Contests is one example, where, for instance, programming competitions have been regularly arranged. These are, however, aimed at talented students and are as such not a suitable channel for introducing computing to students at a larger scale. To mend the situation, Bebras (beaver in Lithuanian) was initiated in Lithuania in 2004 as a contest suitable for all children and youth aged 8-19, inviting them to work with motivating and playful tasks related to computing and computational thinking. The contest is organised online and participants are divided into five age groups (Mini, Benjamin, Cadet, Junior and Senior). Over the years Bebras has grown into a large annually arranged international contest, having about 900 000 participants from 35 countries in 2014.

Sweden has organised the contest twice in addition to a test round in 2012. In 2014 the contest had 18 problems in each category that should be solved in 40 minutes. In 2013 there were 15 problems in each category that should solved in 45 minutes. The reason for the change was to use both the same problems and the same rules as Germany, Lithuania and Finland in order to be able to compare the results between these countries. Unfortunately we had to replace one of the German tasks since it had already been used by Sweden and Finland earlier. Based on the results from the 2014 contest the best boys and girls in each category was invited to Linköping University to participate in an onsite final. The event was greatly appreciated by the participants even though they felt that it was not necessary to have another contest, especially not those that had won their categories since they could only do worse. In its short existence, the participation has grown dramatically from 1869 in 2013 to 7059 in 2014. To our delight, the number of girls participating has also increased from 37% in 2013 (695 out of 1869) to 44% in 2014 (3126 out of 7059). The detailed figures for 2014 can be found in Table 3.

Category	Participants	Teachers	Schools	Cities	Boys	Girls
Mini	1148	61	42	37	565	583
Benjamin	1499	54	51	41	767	732
Cadet	2045	62	60	43	1116	929
Junior	1701	39	37	31	924	777
Senior	666	22	22	20	561	105
TOTAL	7059	189	150	92	3933	3126

4   Activities in Linköping

Since 2013 the Department of Computer Science at Linköping University (IDA) has been actively involved in supporting computational thinking at all levels of the education system.

CoderDojo: To make programming available to as many people as possible, IDA has been active in starting a voluntary programming club based on the international Coder-Dojo concept (coderdojo.org). This is a fun and creative way of supporting kids between 7 and 17 in learning to program. The basic setup is to invite kids to learn to program supported by mentors. A mentor is someone with some experience of programming. The kids are given some hints on what they can do and then they use their creativity to find their own way to learning programming with some nudges and help from the mentors. It is not a class and we try to limit the similarities with school activities. We have encouraged everyone to start with the Hour of Code (code.org) and then explore programming through the use of Scratch [16], a visual programming language from MIT. More advanced kids have started with JavaScript using CrunchZilla. The first CoderDojo in Sweden was started in 2012 in Malmö. CoderDojo Linköping has arranged CoderDojos every second week since January 2014, with about 20-30 kids each time. Most kids are around 7-10 years old. Among those about 60% are boys and 40% girls. Among the older kids almost all are boys. In the spring of 2015 a national organization, CoderDojo Sweden, was started which supports the growing number of CoderDojos around Sweden. There are currently about 15 active CoderDojos in Sweden.

Pupil and Teacher Workshops: To encourage schools that arrange local programming profiles or have pupils that are interested in programmering IDA has arranged programmering a number of half-day workshops. These workshops usually start with an introductory lecture, followed by two lab exercises one involving programming the Nao humanoid robot using the graphical programming language Choregraph and one involving programming generative art in JavaScript, the workshop is concluded with a popular science presentation of research in AI and robotics. We have arranged roughly 15 workshops the last 18 months. We have also arranged a number of workshops to educate and inspire teachers to start using computational thinking in their classes. This also included two special events, one for politicians in Linköping and one for the general public at the big political event Almedalen.

Computational Thinking in Swedish Basic Education: is a project whose purpose is to propose an approach to introducing computational thinking in Swedish basic education and to develop teaching materials for introducing principles and practices from computing as well as computational thinking in a variety of existing subjects (Swedish, Math, Sciences, and Technology). The project involves both researchers at IDA and teachers from Linköping schools and is funded by VINNOVA.

The aims of the project are to:

- write a scientific report on computational thinking and how it can be introduced in Swedish schools,
- develop lesson plans and activities together with teachers, which can be used to introduce computational thinking in different subjects,
- organise teacher training on computational thinking and related didactical aspects, and
- empirically evaluate the lesson plans and activities together with pilot teachers.

As we see it, introducing computational thinking at lower levels of education calls for a dialogue between teachers, teacher trainers and researchers in computer science education. As researchers, we do not have enough insight into the everyday practice in schools nor the long experience of teaching children and youth that teachers have. On the other hand, since computing is not part of the curriculum, most teachers do not have any background in computational thinking or programming, and hence need support in order to understand what computational thinking is, why it is important, and how it can be integrated in their teaching practice. Through the steps above, in particular the three latter ones, we aim at building a model where teachers bring their own expertise in terms of, for instance, subject knowledge and teaching experience, whereas researchers in computer science education can show how principles and ideas from computing can be used in a relevant way in various subjects. The goal is for the model to empower teachers through a continuous dialogue and concrete collaboration.

Discussion. The experience from the project has so far been positive. Teachers appreciate that we take their practice and their students' learning (what is in it for them?) as the starting point, not a given technology or tool (what can we do with x?), which is commonly the case when talking about IT at schools. Based on our discussions with teachers, computational thinking seems to be a more suitable concept to start with than programming or computer science, which are considered both too technical and narrow. In the pilot study 10 teachers were involved. They participated in three half-day workshops and did at least one activity related to computational thinking with computers and one without (unplugged) in their classes. In the first workshop the concept of computational thinking was introduced. In the second workshop concrete activities in the teacher's subjects were discussed and each teacher committed to performing one activity with computers and one without. In the third and final workshop the teachers reported their experience from running the activities. This showed that it was possible for teachers from grade 2 to grade 9 to perform activities related to computational thinking within their subjects with very limited training.

As a direct consequence of the project, IDA and the city of Linköping are discussing ways to extend the collaboration and introduce computational thinking to a much wider selection of schools.

5   Activities in Lund

Lund University is engaged in activities related to computing at school through its science center [6]. The general goal of the science center is to reach out to pre-university education and help to increase the interest in engineering education among the youth in southern Sweden. The science center opened in September 2009 and has since then had more than 166 000 visitors of all ages (as of January 2015). Visitors during weekdays range from school groups to company events and training. The center also takes bookings for a variety of different events and celebrations. It is open to the public on weekends and school holidays, when everyone is invited to try out interactive experiments, attend a show, or see an exhibition.

Before 2012, our science center experiments focused on areas such as physics, chemistry, and electronics, but the area of computer science was lacking, and to fill this gap, a project called "Programming for Everybody" (subsequently denoted PfE) started in 2012 funded by the Engineering faculty and hosted by the science center at Lund University with project members from the department of Computer Science. The main goals of the PfE project are to: (1) develop programming experiments for visitors with particular focus on groups visits from schools, and (2) to develop teacher training so that pre-university schools for all ages can help young learners to discover the excitement and importance of computer programming.

Development of an Open Pedagogical Concept. Within the PfE project, a pedagogical concept for teaching programming is developed. The concept is targeting young learners of age 7 and upwards as well as their school teachers, with the only pre-requisite of being the ability to read from a computer screen and use a keyboard and mouse. When selecting a platform for the pedagogical concept, including language and integrated development environment, these criteria were stipulated:

1. The platform should be free to install, available as open source, and run on Linux, Windows and Mac.
2. The user interface should be available in Swedish, and the programing language should allow Swedish identifiers in the code.
3. The platform should offer 'real-world coding', i.e. it should be based on a modern and professionally used programming language that provides access to general-purpose code libraries.

Based on these criteria, the programming environment Kojo [4] and the programming language Scala [9] was chosen. Both Kojo and Scala are free and open source. The Kojo project is lead by Lalit Pant at Kogics in India [13] and the Scala project is lead by Prof. Martin Odersky at EPFL in Switzerland [14]. Prof. Björn Regnell at Lund University has contributed with translations of Kojo's turtle graphics API and graphical user interface to Swedish, as well as software development, testing, and project sponsoring.

The PfE pedagogical concept includes a series of *programming challenges* [12] that cover a progression of programming concepts including sequential execution, repetition, abstraction, parametrized abstraction, and nested abstractions. The PfE challenges have been iteratively developed based on feedback from kids and teachers.[1]

The PfE programming challenges are rooted in a contructionist approach [10] through a Swedish turtle graphics API in Scala. The initial challenges are based on reading and tweaking a given, worked example with a code snippet and hints, to enable learners to quickly grasp syntax as well as concepts, in line with suggestions that worked examples may (according to cognitive load theory) be effective in learning how to program [5].

Programming Activities for Science Centre Visitors. Since the PfE project started, we have had more than 10000 young learners have experienced our programming experiments, based on the PfE challenges with Kojo and Scala. The visits are mainly school classes that try out different experiments in the science centre, where programming is one station. The school classes are divided into smaller groups of around 10 students, often working in pairs to solve the PfE programming challenges [12], while one or two instructors are providing guidance. School class group session range from 20 to 45 minutes and focus on drawing pictures with turtle graphics, while learning about sequence, repetition and abstraction using procedures without and with parameters.

Programming Education for Teachers. More than 100 teachers have passed our programming courses comprising 2-3 half-days with assignments in between, to try out programming in class using Scala and Kojo. The teachers learn programming with the help of the PfE programming challenges, which they then try in their classes with young learners. Teachers then share their experiences with each other, including new challenges that they develop in relation to their subject curricula.

---

[1] The challenges are available with a Creative Commons Attribution-NonCommercial-ShareAlike licence [12].

Discussion. Teachers that attend our courses are teaching different subjects at primary school, ranging from maths and science to language and sports, but also practical skills such as needle- and woodwork. Many teachers that attend our courses are at first uncertain if they will manage to run a programming class within their subject, but when they experience how their students engage in the challenges and with enthusiasm discover the playfulness and joy involved in creating software, many change their mind and see the opportunities. A major opportunity for the future is to create an active network of primary school teachers to enable sharing and inter-scholar learning of how to integrate programming in the existing curricula. The teachers that have taken our courses at the LTH Science Centre realize the potential, but they want more subject-specific study material and also assessment models so that the student learning outcome from programming projects can be assessed in relation to the learning goals of their subject.

6 Activities in Luleå

Starting in 2013 a collaboration between Luleå University of Technology, LTU and The Municipality of Luleå [2] started around how to increase awareness of computational thinking and digital literacy in Luleå's schools. The collaboration started informally and was made more formal when funding was secured during 2014 for several projects. It was early identified that no single solution would be enough to inspire and educate the more than around 1900 teachers and principals but rather that several different activities had to be organised and implemented. It was also identified that the best approach was to test different activities to see which were the most appreciated. The activities organised include *Pedagogical Pubs with TeachMeet*, *Open Educational Workshops*, *Programming inspiration* and *Making in Schools*.

Pedagogical Pubs with TeachMeet: In an effort to inspire teachers and allow teachers to share their ideas, thoughts and progress several Pedagogical Pubs with TeachMeet were held during 2014 where invited speakers presented various topics related to ICT in Schools and followed by very short fire talks by teachers that wanted to share what they had done in their classrooms. Each of these gatherings had around 80-120 participants and they were open to the general public as well including teacher students from the university.

Open Educational Workshops – CS4HS Luleå 2014: The Open Educational Workshops were supported by Google via their CS4HS program during 2014. The goal of these workshops was to give teachers and principals a chance to meet and learn more about Computer Science and how it can be applied in grades 1-12. The workshops were focussed on giving concrete hands-on work with various tools including graphical programming like Scratch and Blockly via Hour of Code, testing electronics like Arduino and doing some more advanced Javascript programming. 5 workshops were held of which the first 4 were with computers while the final one was *unplugged* where the participants discussed how to apply their new knowledge in the classroom and also included a specific discussion on how to better include young females.

---

[2] Many of the efforts described here were organised together with principal Agneta Hedenström from the Luleå Schools Antnäs and Måttsund.

The workshops were held free of charge and over 80 persons applied of which 35 participants were selected based on geographical location as well as teaching level. The goal was to get a good mix between schools and teacher levels. The participants showed a real interest and really wanted to learn and also learn how to apply their skills in their classrooms and it was very rewarding to see how they immediately applied them the after the workshop was held.

Programming Inspiration – EU Code Week, Hour of Code and School Visits: During the autumn of 2014 two separate events around programming in schools were held. In October the EU Code Week was held and in Luleå 9 separate events were held (out of 90 in Sweden) where students got to try Scratch primarily at grades 2-6. The EU Code Week was held all over EU at the same time with more than 3000 events during one week. During December 2014 the global Hour of Code week was held and in Luleå the efforts were focussed to December 9 when with the help of volunteers we were able to reach more than 1000 students via more than 50 separate programming hours where students got to try the Hour of Code learning environment or Scratch. Besides these larger organised events a number of schools visits were held where programming and making was presented to students at various levels in the school system.

Making in Schools – Luleå Makerspace and the Skaepiedidh Project: Another approach has been through *Making* in schools where the students have gotten to test various technologies. This has mostly been done through collaboration with the Luleå Makerspace which is a non-profit organization with the goal to promote ICT and technology interest to persons of all ages. It was early identified that the teachers only needed a small amount of *pushing* in the right direction to get started with making and programming in school, but at the same it was hard to get a scalable solution as much relied on personal contact to inspire them. Thus, the Skaepiedidh project was created between LTU and the Luleå Municipality with national funding from VINNOVA with the goal to create an online system for doing the inspiration online where teachers could exchange recipes for making and programming in schools, comment on these recipes, remix them and spread their implementations further. More thoughts on Making in schools can be found in [11].

Discussion. The various efforts presented here have been very appreciated by those participating which is shown via that the teachers actually get started with computational thinking in schools as well as that they want more help and more inspiration. The Luleå Model is a combination of several different efforts as to reach as many as possible no single effort is enough. At the same time it has been identified that only a smaller portion of the 1900 teachers and principals are reached and some of the schools are not present at all and thus in turn a large portion of students are not reached unless a more formal mandate by the government and/or the schools leaders is made. During 2015, the efforts will continue but with more of a focus trying new things as well as influencing politicians and school leaders through local and national efforts. Several workshops and a larger educational conference is planned as well with the goal to spread good examples and get people to meet and talk and learn from each other as we foresee peer learning as one of the most important parts of this campaign.

7   Activities in Stockholm

The Stockholm region is a large area, with over two million inhabitants, which makes collaboration around topics, like computational thinking, challenging to coordinate. Within the region, some municipalities have been more progressive about getting their teachers to start using computational thinking in the classroom, even though still in small scale. During 2014 computing and programming became popular buzzwords. Consequently, programming has been introduced at several schools in the region by passionate teachers. In February 2015, people representing the City of Stockholm, the Stockholm Chambers of Commerce and the company Spotify, wrote an article in Dagens Nyheter about a larger effort to set up pilot schools to teach programming in elementary schools in Stockholm. This is now under preparation. Teachers and principals from schools already working with computational thinking, are consulted to find best practice in the implementation to all 265 schools within the City of Stockholm. This is so far the largest intervention in Sweden and will be of big importance as an example for the rest of the country.

Programming Inspiration – EU Code Week, Hour of code: During the 2014 European code week, Academedia, the largest group of independent schools in Sweden, had students from all secondary schools teach 5th graders about programming. At least 20 schools within the Stockholm area participated. The documentation of participants haven't been that precisely. This is something to improve this coming year. As a total 10 schools within the area have reported participation during Hour of Code in December.

Organizations:

- Coder Dojo was the first code initiative in Stockholm in 2012. Two days a week, CoderDojo offers open programming sessions for kids, at a library in the city center, free of charge and there is no need to bring a computer.
- The National Computer Society, Dataföreningen, has organized several events for children and also teachers and other adults interested in coding with Scratch. The goal is to use the members of the society as mentors to the teachers who want to try coding in class. Dataföreningen also has RaspberryPis to bring to schools if they don't have enough computers. There are two networks within Dataföreningen, one for anyone interested in contributing to children learning more about programming, and one for teachers.
- Kosmosklubben, an NGO offering 10 different programming courses every week and 9 classes of robotics. There is also a special program for high school girls from suburbs, to get them more interested in code and IT.
- Kids Hack Day, is an after school program, where kids can pay to participate in coding and maker activities. They have had three groups during the semester. The cost is rather high but includes all kinds of maker supplies.
- Kodcentrum: Is a non-profit organization based in Stockholm offering children and youth hands-on experience in programming through an after school program. This is in contrast to CoderDojo more like a course were the children follow a prepared material during a whole semester. Kodcentrum has activities in Göteborg, Linköping, Stockholm, Sundbyberg, Umeå and Uppsala.

Tekniska Museet (Museum of Technology), organizes children's hack sever times during the year. They have both courses for beginners and more advanced programmers. Tha hack club uses Scratch, but the museum also offers robot programming with LEGO Mindstorms. In September a new part of the museum will open, called the Mega mind, with one maker space and one coding space.

Initiatives for girls: To try to get more girls interested in engineering and IT, separate groups are more and more popular. Here are some initiatives.

- Geek Girl Mini, an after school program for girls in 5th grade. Two schools in Stockholm has been involved. One in an upper middle class-area and one in an all immigrant area. The second group was the largest and most of the 21 girls showed up every time, even on holidays. It was a big success.
- Girls Code, after school program at Mälarhöjden for girls in grades 7-9.
- Tech Girl, an initiative from the company Valtech. Two groups of girls in 6th grade have been taught programming in scratch for one semester each. Volunteers from Valtech have been running the workshops.
- MakerTjej, an organization with a mission to get more girls into the maker movement.
- Tjejhack, an organization with a mission to get more girls into the gaming business and also to change the rather raw climate in the gaming community.
- Teklafestivalen, a festival about music and tech, organized by the Royal Institute of Technology and the artist Robyn. The evens was a huge success, with over 2000 applications for the 200 tickets. The participants were chosen to get a variation in age and geographic background. Companies as Valtech, Google and Spotify arranged workshops during the day. Hopefully the event will be annual.

Teacherhack: Since programming is not part of the national curriculum, a group of teachers started to search for ways to integrate different aspects of computational thinking and computing in every subject within the current curriculum. On the website teacherhack.com there are texts explaining all 20 subjects in the curriculum from a digital aspect, over 30 different lessons and comments from around 30 teachers representing different subjects. The website is not yet officially launched, but already it has some 200 visitors each day. The goal is for Teacherhack to become a valuable resource for teachers wanting to include computational thinking and programming in their classrooms.

Sjöstadsskolan: At Sjöstadsskolan in Stockholm, programming has been taught since February 2013. During the first year, programming was introduced only to 50 students in 5th grade, but 2014-2015 all 800 students, including preschoolers (from 3 years old) have tried programming or computational thinking in some form.

- Preschool (3-6 years): Bee-Bots for creating simple programs after step-by-step instructions.
- 1st grade: Various games for understanding code "unplugged". iPad apps, such as BeeBot and Kodable. Code.org.
- 2nd grade: Games for understanding binary code and the Internet. Bee-Bot programming. Participated in Bebras.

- 3rd grade: Code.org and creating mazes for the BeeBot using paper, pen and ruler.
- 4th grade: Investigating geometry through the programming environment Kojo. Constructing real-life computer games in the school yard.
- 5th grade: Participating in a study on teaching Cartesian coordinates, comparing traditional instruction to introducing coordinates using Kojo. Participated in Bebras.
- 6th grade: Students already had 1.5 years of programming experience. Programming as an extracurricular activity.
- 7th grade: Code.org. Participated in Bebras.
- 8th grade: Code.org. Students learned how to use block based programming for creating visualizations in biology. LEGO Mindstorms were used in Technology class.
- 9th grade: Code.org.
- Special need school: Two students (aged 6 and 16) with severe autism have been practicing with BeeBot.

Research: Sjöstadsskolan has conducted three pilot studies about programming in school. The first pilot study was conducted on introducing computational thinking through programming in a Swedish course [8]. The initial motivation was to investigate the use of programming when teaching 4th graders (9-10 year olds) how to write instructions and reflections in an authentic way. The children were given assignments to be solved in pairs or small groups. A shared blog was used to distribute the results, which also made it possible to discuss and compare solutions. The same methodology was used in the 5th grade as well (10-11 year olds), then including more aspects of computational thinking. The second study was conducted with computational thinking as a separate subject. The conclusion was that the students found it fun and interesting, but not as part of the rest of the school day. The transfer from computing to the other lessons, just didn't happen. The third study, done in cooperation with Stockholm University and City of Stockholm department of education, was focused on using programming as a tool in teaching math. The study focused on Cartesian coordinates and the result will be presented at a conference for teacher's research in November 2015.

Discussion: Based on the initial pilot study and discussions afterwards we identified several ways in which computational thinking can be used in language courses. First, programming languages are also languages, sharing for instance the notion of grammar (syntax) and meaning (semantics); hence these can be discussed in traditional language courses. Moreover, computational thinking concepts can be introduced in a natural way when talking about languages. For instance, abstraction through nouns and verbs etc., problem decomposition through a grammar that breaks down the problem of describing classes of sentences into smaller problems; patterns through rules in a grammar; programming languages used for implementing rules in a grammar; introduction of algorithms and instructions through conditions, loops and iteration used in, for instance, storytelling. The second study implies that it's not for the best to teach computational thinking as a separate subject. To better use the transfer effect, we should integrate it in other subjects. The result from study number three, are yet not presented, but it seems to be a reasonable idea to use a program environment like Kojo, to teach Cartesian coordinates. As the capital and the largest city in Sweden, the path that the City of Stockholm will choose, will probably be of big importance for the implementation of computational thinking in schools.

8   Lessons Learned and Ways Forward

All the initiatives described above show that computational thinking and programming can be introduced and support learning at lower levels of education, although such elements are not part of the curriculum. There is nothing in the national curriculum that prevents schools and teachers from working with computational thinking, but also nothing that encourages them to do so. Since schools and teachers are already under a lot of pressure from many directions, few do anything. Those that do, however do very good things. In order for these initiatives to spread and become commonplace throughout the nation, there is a need for a national initiative to introduce computational thinking in the school system. There are indications that such initiatives might be coming.

Another major challenge is teacher training. Teacher education programmes do not commonly include computing or computational thinking in their curriculum and consequently pre-service and in-service teachers lack insight and skills in these areas. We are therefore actively seeking a dialogue with the universities providing teacher education, but so far they have showed little interest. This calls for a dialogue between teachers, teacher trainers and researchers in computer science education.

Based on our current activities and lessons learned, we are currently working towards the following goals:

1. Establishing the term "Datalogiskt tänkande" as the Swedish term for computational thinking, implying the general skill set that can be trained through programming and that can be used in an interdisciplinary way throughout the curriculum.
2. Engaging as many schools as possible in Bebras, in order to stimulate the interest for computational thinking.
3. Supporting informal activities such as CoderDojos,and Maker Spaces which play an important role in giving students hands-on experience with, for instance, programming.
4. Collaborating with municipalities wanting to introdue or at least test computational thinking at a larger scale.
5. Supporting teachers in developing concrete example activities and lesson plans on introducing different aspects of computational thinking in a variety of subjects.
6. Designing concrete suggestions for professional development for teachers on computational thinking, for instance, in the form of a nation-wide MOOC supported by local study groups.
7. Engaging in continuous discussions with teacher education programmes in order to introduce at least one compulsory course on computational thinking for all pre-service teachers.
8. Developing suitable means for assessing computational thinking, for instance based on Bebras activities.

We have in this paper only reported activities that we are directly involved, but there are several other initiatives in Sweden that focus on the introduction of programming and computer science to young learners. Given the awakening interest from politicians and education developers, we believe that computational thinking soon will play a more prominent role in Swedish pre-university education.

References

1. ACM Inroads (December 2014)
2. Barr, V., Stephenson, C.: Bringing computational thinking to k–12: What is involved and what is the role of the computer science education community. ACM Inroads 2(1), 48–54 (2011)
3. Digitaliseringskommissionen. En digital agenda i människans tjänst - en ljusnande framtid kan bli vår (2014)
4. Kojo home page, http://www.kogics.net/kojo
5. Lister, R.: After the gold rush: Toward sustainable scholarship in computing. In: Proceedings of the Tenth Conference on Australasian Computing Education, ACE 2008, vol. 78, pp. 3–17. Australian Computer Society, Inc. (2008)
6. Lund University Science Center home page, http://www.vattenhallen.lth.se/
7. Mannila, L., Dagiene, V., Demo, B., Grgurina, N., Mirolo, C., Rolandsson, L., Settle, A.: Computational thinking in k-9 education. In: Proceedings of the Working Group Reports of the 2014 on Innovation & Technology in Computer Science Education Conference, ITiCSE-WGR 2014, pp. 1–29. ACM, New York (2014)
8. Nygårds, K., Mannila, L., Heintz, F.: Computational thinking in teaching Swedish in the middle school. In: ICED 2014: Educational Development in a Changing World (June 2014)
9. Odersky, M., Altherr, P., Cremet, V., Emir, B., Maneth, S., Micheloud, S., Mihaylov, N., Schinz, M., Stenman, E., Zenger, M.: An overview of the Scala programming language. Technical Report, IC/2004/64, EPFL Lausanne, Switzerland (2004)
10. Papert, S.: Mindstorms: Children, computers, and powerful ideas (1980)
11. Parnes, P.: Skapande och skaparkultur som drivkraft för kreativt lärande i skolan (February 2015)
12. Björn, R. (ed.): Programming Challenges, http://fileadmin.cs.lth.se/cs/Personal/BjornRegnell/uppdrag.pdf
13. Regnell, B., Pant, L.: Teaching programming to young learners using scala and kojo. In: LTHs Pedagogiska Inspirationskonferens, vol. 8, p. 4. Lund University (2014)
14. Scala home page, http://scala-lang.org/
15. European Schoolnet. Computing our future. computer programming and coding - priorities, school curricula and initiatives across europe (October 2014)
16. Scratch home page, http://scratch.mit.edu/
17. Wing, J.M.: Computational thinking. Communications of the ACM 49(3), 33–35 (2006)

A Snapshot of the First Implementation of Bebras International Informatics Contest in Turkey

Filiz Kalelioğlu[1], Yasemin Gülbahar[2], and Orçun Madran[3]

[1] Başkent University, Department of Computer Education and Instructional Technologies,
Ankara, Turkey
filizk@baskent.edu.tr
[2] Ankara University, Department of Informatics, Ankara, Turkey
gulbahar@ankara.edu.tr
[3] Hacettepe University, Department of Information Management, Ankara, Turkey
omadran@hacettepe.edu.tr

Abstract. Computing was perceived as an essential skill for computer scientists, engineers, mathematicians and those from similar disciplines. Today, to the contrary, people of most ages are expected to possess basic computing skills in parallel with the requirements of up-to-date technological tools. To equip students with the necessary skills, computer science courses need to be delivered compulsorily, or at least delivered as a part of another course for almost all age groups and levels. Besides delivering these courses, awareness of this valuable skill is also essential, and for this aim, Olympiads or contests are now held in many countries. Bebras International Contest is one such organisation. In December 2014, Turkey also participated in this contest with 1,788 elementary students from different cities. This paper examines the student performance of the 2014 Bilge Kunduz (the Turkish term for Bebras) International Informatics Contest and explores coordinators' perceptions about the contest. Based on the student performance and overall success, the average score in Turkey was 65.01 (where scores varied between 0 and 135). According to the perceptions of coordinators, it can be said that the contest was favoured by all coordinators, but that some enhancements to the contest platform are needed.

Keywords: country report, Bebras contest, students' performance, coordinator perceptions.

1   Introduction

Since its first invention, computing was perceived as an essential skill for computer scientists, engineers, mathematicians and those from similar disciplines. Today, to the contrary, people of almost all ages are expected to possess basic computing skills, in parallel with the requirements of up-to-date technological tools. Thus, the discipline of informatics is gaining popularity and the teaching of it also becoming more important since people are expected to solve real life problems, think critically and computationally, be creative, make decisions and be aware of technological tools and concepts as digital citizens [1, 2].

© Springer International Publishing Switzerland 2015
A. Brodnik and J. Vahrenhold (Eds.): ISSEP 2015, LNCS 9378, pp. 131–140, 2015.
DOI: 10.1007/978-3-319-25396-1_12

For our students to be equipped with the necessary skills, computer science courses need to be delivered compulsorily, or at least delivered as a part of another course for almost all age groups and levels. Besides delivering these courses, it is important to make the public aware of Informatics, not only as a technology, but also as the science to educate them and improve their experience with technology. Organising an informatics contest is one way of achieving this goal, hence in 2014, Turkey participated in the Bebras International Informatics Contest.

Having been held in different countries since 2004, Bilge Kunduz (the Turkish term for Bebras) International Informatics Contest was organised in Turkey for the first time in 2014. The competition will continue to be held in November each year. This research study has been conducted to reveal the process and experiences of this first pilot implementation in Turkey.

2   Method

This research is a case study composed of quantitative and qualitative measures since it is specific for a country. Case study is an in-depth exploration of a bounded system such as activity, event, process, or individuals based on extensive data [3]. The case for this study is the implementation of Bebras international informatics contest in Turkey. Hence, all aspects of the contest have tried to be investigated, both from the perspectives of students and school coordinators. This research study seeks answers to the following research questions:

1. What was the overall success of the students?
2. What was the success according to the degree of difficulty of the questions?
3. What was the success according to the sub-fields?
4. What were the detailed item analyses of each question?
5. What were the thoughts of the school coordinators in terms of application of the contest?

    a. What do they think about the Bilge Kunduz International Informatics Contest?
    b. What were the reactions of their students about the competition process?
    c. What were the problems faced during the competition process?
    d. What do they think about Moodle platform which hosted the competition?
    e. What do they think about the Bilge Kunduz tasks presented in the competition?
    f. What are their suggestions to improve the effectiveness of the competition?

2.1   Participants

Schools & Students

Bilge Kunduz International Informatics Contest was carried out as a pilot in the cities of Ankara, Izmir, Erzurum, Samsun, Adana, Muğla, Istanbul and Çanakkale in the 3rd week of December, 2014. A total of 57 schools participated in the contest on a voluntary basis, with the support of 12 provincial coordinators, and 1,788 students successfully competed in the contest. Of the 57 schools that participated in the contest, 31 (54.39%)

were public schools and 26 (45.61%) were private schools. In addition to the pilot cities, Aksaray, Niğde, Tekirdağ and Sivas also participated in the contest.

School Coordinators

In order to better understand the contest and to understand the problems they faced, opinions of the school coordinators (as opposed to the provincial coordinators) were gathered regarding their experiences as coordinators and about the process of the competition. Eight open-ended questions were given to the coordinators, with responses received from 20 of them. The professional teaching experience of the coordinators ranged from one to 13 years.

2.2   Context and Procedure

Questions for the Bilge Kunduz International Informatics Contest were selected from questions prepared in a workshop by representatives from more than 30 countries. According to the procedure, questions are prepared according to predetermined criteria [4]. Then, each country selects questions from two separate question pools (compulsory questions and suggested questions), and implements them after translating into the preferred language. The selection process is very important, and questions must be chosen according to sub-domains within the scope of informatics. Moreover, [5] explain the sub-domains as in table 1.

Table 1. The sub-domains of the tasks

Sub-domains	Explanations
INF	Information comprehension, representation (symbolic, numerical, visual), coding, encryption
ALG	Algorithmic thinking including programming aspects
USE	Using computer systems (e.g. search engines, email, spreadsheets, etc.) General principles, but no specific systems
STRUC	Structures, patterns and arrangements, combinatorics, discrete structures (graphs, etc.)
PUZ	Puzzles, logical puzzles, games (Mastermind, Minesweeper, etc.)
SOC	ICT and society, social, ethical, cultural, international, legal issues

Problems must be chosen in order to include all sub-fields. Questions should be interesting for the students, should motivate learning, and should allow students to demonstrate their knowledge and skills. In this context, the contest was carried out by selecting five easy, five medium and five hard questions, including various sub-domains, from the international question pool. Subdomains of selected questions and the questions are shown in the table 2.

Table 2. Subdomains of selected questions

	ALG	INF	STRUC	PUZ	SOC	USE
Easy[1]	3	2	-	-	-	1
Medium[2]	3	2	2	2	-	-
Hard[3]	4	2	1	3	-	-

[1] Non-ordered stars, Phone keyboard, Ice cream, Lost in a City and Assemble the Fish
[2] Abacus, Truchet, Family Tree, Broken Clock and Stairs Robot Snake
[3] Cutting down trees, Bridges, Price of a gift, Truth and Bagels

2.3 Data Collection and Analysis

Data needed to undertake detailed analysis about the questions and the student's success were gathered from the platform upon which the contest was carried out (Moodle). To obtain coordinators' perceptions, an online questionnaire composed of seven questions was sent out via e-mail. The first part was analysed quantitatively based on descriptive statistical measures, whereas the open-ended questions were analysed through content analysis. Moreover, significant ideas and statements by some of the participants are included as quotations to illustrate the findings.

3   Results

3.1 Overall Success

The graph below shows the overall success in percentage terms of the 1,788 fifth and sixth grade students who participated in the contest. No students answered all questions correctly. Students remained in the contest from between two minutes and 45 minutes.

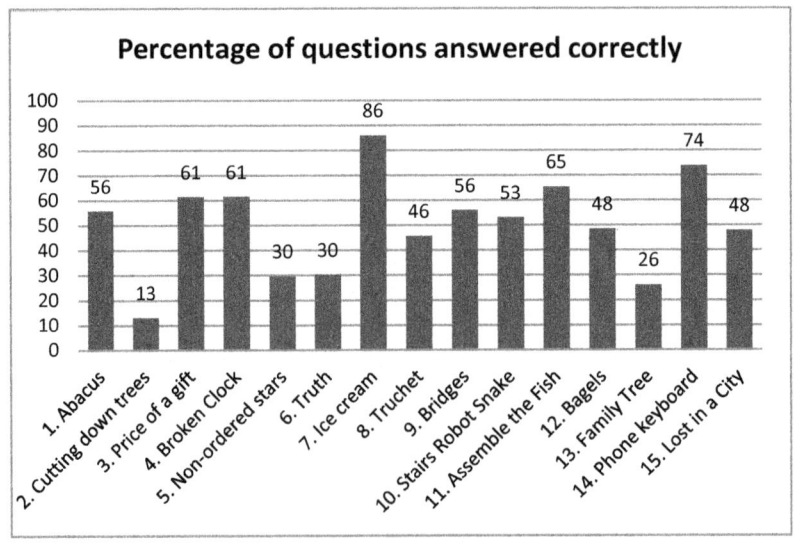

Fig. 1. Percentage of questions answered correctly

The scores varied between 0 and 135. The average score was 65.01 and standard deviation was found to be 26.05. The overall success of students in terms of scores achieved (the highest score being 129) are shown in the following graph.

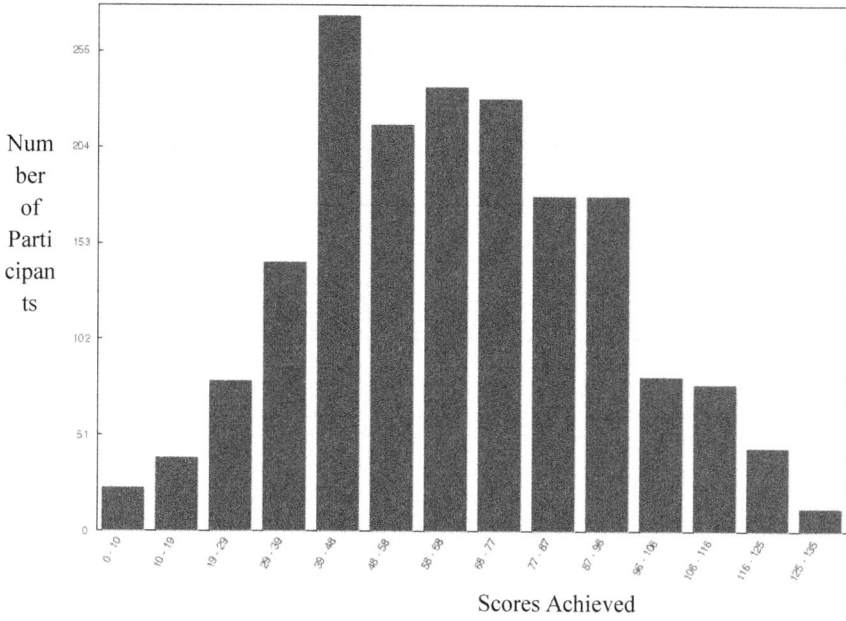

Fig. 2. The overall success of students

3.2   Success According to the Degree of Difficulty of Questions

Success based on the degree of question difficulty is shown in the table 3. As expected, students' success decreased as questions get more difficult.

Table 3. Success based on the degree of question difficulty

Easy	60.50%
Medium	48.38%
Hard	41.81%

3.3   Success According to the Sub-Fields

As seen, the sub-field where students were the most successful is algorithmic thinking. Appropriate ways and methods will continue to be investigated in order to improve other sub-fields.

Table 4. Success based on the degree of question difficulty

Sub-field	Number of questions	Average score
ALG	10	40.88
INF	6	27.74
PUZ	5	19.89
STRUC	3	13.17
USE	1	3.93

3.4   Item Analysis

Detailed analysis of each question is shown in the table 5.

Table 5. Detailed analysis of each question

	Facility Index	Std. Dev	Random guess score	Intended weight	Effective weight	Discrimination index	Discriminative efficiency
1. Abacus	55.80%	49.68%	25.00%	7.00%	7.96%	37.04%	46.24%
2. Cutting down trees	13.21%	33.87%	25.00%	9.00%	3.63%	-3.95%	-5.90%
3. Price of a gift	61.30%	48.72%	25.00%	9.00%	8.58%	25.20%	32.09%
4. Broken Clock	61.63%	48.64%	25.00%	7.00%	7.66%	34.13%	43.55%
5. Non-ordered stars	29.71%	45.71%	25.00%	4.00%	4.87%	21.37%	27.04%
6. Truth	30.15%	45.90%	25.00%	9.00%	8.80%	32.67%	41.86%
7. Ice cream	86.01%	34.70%	25.00%	4.00%	4.56%	28.98%	49.19%
8. Truchet	45.64%	49.82%	25.00%	7.00%	7.74%	33.56%	41.20%
9. Bridges	55.80%	49.68%	25.00%	9.00%	9.53%	36.09%	45.03%
10. Stairs Robot Snake	52.92%	49.93%	25.00%	7.00%	7.88%	35.40%	43.81%
11. Assemble the Fish	65.41%	47.58%	25.00%	4.00%	5.10%	22.77%	29.52%
12. Bagels	48.53%	49.99%	25.00%	9.00%	7.74%	13.91%	16.86%
13. Family Tree	25.99%	43.87%	25.00%	7.00%	4.49%	3.33%	4.25%
14. Phone keyboard	73.96%	43.90%	25.00%	4.00%	5.71%	36.18%	50.55%
15. Lost in a City	47.75%	49.96%	25.00%	4.00%	5.78%	30.16%	36.83%

3.5   Perceptions on Bebras International Informatics Contest

All coordinators were positive about the Bebras Contest. Their perceptions were grouped under two themes, namely: "the purpose and quality of the contest" and "Bebras tasks". Related to the purpose and quality of the contest, it was obvious that the contest was perceived as adding value to the national course on the topic of "ICT and Software", supporting students learning by entertainment. On this topic, one coordinator said "We were happy that we have a contest about informatics" whereas another stated "It is a very useful contest, especially for disseminating the national course and revealing the necessity of teaching algorithms. It was the first attempt where the leading role belongs to the ICT course".

While another coordinator explained "I think that this was a good activity which motivated students. It may turn to Olympiads and became the prestigious indicator for our course" and another added "It was a quite different experience for our students. Different from all the assessment approaches, it was difficult, but a very entertaining activity". Most coordinators expressed praise and hopes for dissemination across all age groups. About this idea one coordinator said "I wish this contest could be administered at the national level for all schools". Some coordinators expressed satisfaction of their first experience in an electronic environment and with having the contest as Internet-based. One of them revealed "We were very excited to take part in an Internet-based contest. Most of my students are asking when the next contest will be".

In terms of "Bebras tasks", the main perception was the task features of teaching while entertaining. One coordinator said "Tasks can support cognitive development of students and can also enhance their numerical and verbal logic." And another stated that "I observed the positive impact of the contest on my students. I believe that tasks can support students' ways of dealing with real-life problems". They also underlined the quality of the tasks, so that students should need to take care and apply practical intelligence to solve them. One of the coordinators concluded that "The contest is a successful attempt for students to develop their algorithmic and cognitive thinking skills". Hence, it could be stated that the contest is favoured by all coordinators.

3.6   Feedback about the Contest Process

Most of the coordinators stated that students liked, valued and were motivated about the contest. One coordinator said "At first, students were surprised by the logical patterns in the questions, but then they started to like solving the tasks". Most of the coordinators stated that students were impatient to see their scores and kept asking about the announcement date for results. One coordinator said "The first thing they asked about was the gift they would get at the end of the contest. There should be something for encouragement at this age. They found the tasks easy, but they could not match the topics with our ICT course".

Although very few in number, some coordinators stated that some students found the questions very difficult, there was a lack of time and for these reasons, they became bored. Some students were expecting their names to be announced on their school website due to their success, whereas some stated that there should be a threshold for receiving a certificate. For this reason, in future, sponsors could be found to provide small gifts for those who were in the top five or so. This issue can be dealt with next year.

3.7   Obstacles Faced during the Contest

Some students had some technical problems during the contest. Among the obstacles mentioned by the coordinators, there were: Internet connection problems; login problems; issues of being in such a contest for the first time with different types of tasks; having out-of-date computers; lacking adequate time to complete questions; and having heard quite late about the announcement of the contest. One coordinator said that "Students were anxious since this contest was their first online experience", and another coordinator added "The main obstacle was 5th grades not familiar with algorithmic logic". In general, there were no real problems, apart from some minor regional issues reported as obstacles.

3.8   Perceptions about the Online Platform

A widely known and used open source learning management system – Moodle – was used to administer the contest. Most of the coordinators stated that they liked the system and found it easy to use. The system was found to have no technical problems, and was simple and successful. One of the coordinators stated "It was easy to login; the interface was clear and navigation was easy for managing the contest for the students". However, some coordinators said that the interface should be changed and that there should be a special template for the contest. On this topic, one coordinator said that a more entertaining interface could be used for younger age groups, and another added that the interface is suitable for adults and should be simplified visually for lower age groups. Thus, the online platform could be enhanced in terms of interface and some aspects of usage in order to fulfil the expectations of coordinators.

3.9   Perceptions about Bebras Tasks

Most coordinators were positive about the Bebras tasks. Some of these ideas are presented below:

- "Tasks were highly qualified and their level was appropriate for the age groups."
- "Although they had difficulties with some questions, tasks were suitable and very logical."
- "Questions were really good and addressed thinking skills."
- "Students used logic to find solutions and this was very useful for my students."
- "The students were not used to these kinds of questions, but it should be this way."
- "The questions have the capability to widen the scope of my students."

Moreover, some coordinators criticised the difficulty level of tasks; they stated that there were some inappropriately categorised items in terms of difficulty levels. One coordinator criticised asking the same questions to both the 5th and 6th grades, stating that "questions should differ for each grade". Yet another coordinator expressed his/her wish of having "more questions where the task content addresses real life problems". In general, tasks were favoured and there were no serious complaints.

3.10   Suggestions for Improving the Contest

Suggestions made by coordinators emerged under four main themes, namely: the quality of tasks, announcement of the contest, duration of the contest, and improvements to the system. About the quality of tasks, several coordinators suggested that different grades should be set different tasks, but the majority were happy with the current situation. It was obvious that the announcement of the contest should be made earlier; this was due to bureaucratic slowness in delivering formal messages to the schools. Although a few students could not manage their time and had difficulty, this was however from just a negligible percentage. As a last point, the interface could be adapted to be made more appropriate for different age groups. Therefore, earlier announcement and improvements in the system should be done before next years' administration of the contest.

4   Conclusion

The purpose of this report was to examine the student performance of the Bilge Kunduz International Informatics Contest held on 15-19 December, 2014, and to explore coordinators perceptions about the contest. Based on the student performance and overall success graphs, it can be said that the students were successful, with an average score of 65.01 (scores varied between 0 and 135). As questions became more difficult, the students' success decreased. Results can be regarded as successful because of the quality and level of difficulty of the questions, as well as the requirements for algorithmic skills to solve them. Detailed item analysis shows that "Family Tree", "Cutting down trees" and "Bagels" questions have particularly low discrimination index. This means that high-performing students would not select the correct answer any more than low-performing students. These items should be revised and statements should be re-examined.

According to the perceptions of coordinators, it can be said that the contest was favoured by all coordinators, but they suggested that next year some gifts could be awarded to those in the top five. They were very few problems faced during the contest, but there could be some enhancements made to the platform the contest is delivered on in terms of some features and the interface. Tasks were also generally favoured by coordinators, but they also underlined the importance of early announcement of the contest.

Although this contest is an attempt to disseminate awareness about informatics, most coordinators insisted to be grouped according to their schools' success at the national level. In general, there is a common understanding about national contests held in Turkey, which groups students or schools according to their success and provides awards to the top three or so. This may be why most schools just entered their best achievers in the contest. Efforts to access all students will continue, but we are also trying to emphasise other non-traditional approaches.

Another point that arose from both coordinators and researchers was disagreement about the level of difficulty of the questions. This phenomenon can be considered a cultural difference, or maybe due to the content delivered by the ICT course. This problem can be resolved by proposing Bebras tasks and participating in the related workshops.

In summary, it was decided to improve the contest platform and interface based on the suggestions of coordinators. In this way, coordinators should be able to better understand the success level of their students and also to be able to access some detailed reports. In addition, next year more age groups will be invited to the contest.

References

1. ISTE. : ISTE Standards for Students. http://www.iste.org/docs/pdfs/20-14_ISTE_Standards-s_PDF.pdf (2007).
2. NRC. : A Framework for K-12 Science Education: Practices, Crosscutting Concepts, and Core Ideas. Washington, D.C.: The National Academies Press (2011).
3. Creswell, J. W. : Qualitative inquiry and research design: Choosing among five approaches (2nd ed.). Thousand Oaks, CA: Sage (2007).
4. Vanicek, J. : Bebras Informatics Contest: Criteria for Good Tasks Revised. In: Gülbahar, Y., Karataş, E. (eds.) ISSEP 2014. LNCS, vol. 8730, pp. 17-28 (2014).
5. Dagienė, V., Futschek, G. : Bebras International Contest on Informatics and Computer Literacy: Criteria for Good Tasks. Informatics Education - Supporting Computational Thinking. Lecture Notes in Computer Science, vol. 5090, pp. 19-30 (2008).

Introducing a New Computer Science Curriculum for All School Levels in Poland

Maciej M. Sysło[1,2] and Anna Beata Kwiatkowska[1]

[1] Faculty of Mathematics and Computer Science, Nicolaus Copernicus University
Chopin str. 12/18, 87-100 Toruń, Poland
syslo@mat.umk.pl, aba@mat.umk.pl
[2] Faculty of Mathematics and Computer Science, University of Wrocław
F. Joliot-Curie str. 15, 50-383 Wrocław, Poland
syslo@ii.uni.wroc.pl

Abstract. The first regular informatics lessons in schools in Poland were organised in the second half of the 1960'. Some of them were devoted to programming a mainframe computer (in Wrocław) and some to theoretical models of computers and computations (in Warsaw). Then, for more than last 30 years of formal informatics education in Poland we have been very successful in keeping informatics (as computer science) as a stand-alone subject and in shaping its curriculum according to high standards of the discipline. In this paper, in Section 1 we first discuss terminology related to computers in education and then report on early history of informatics education in Poland. In Section 2, the present curriculum of informatics subjects is described in details together with some comments on using computational thinking in its implementation. Then, as the main contribution of this paper we introduce in Section 3 a new computer science curriculum for all school levels in Poland. To this end, the existing curricula of informatics subjects have been remodeled, extended (e.g. by adding programming to each level), and unified according to the five Unified aims of learning computing. The new curriculum benefits very much from our prior curricula and experience. Finally we discuss some implementation details, supporting activities, and the road map for a successful introduction of the curriculum to all schools.

Keywords: education, informatics, computer science, algorithmic thinking, computational thinking, programming.

1 Introduction

1.1 Terminology

In education, as in the most of other disciplines and areas of computer applications, the terms: computer science, computing, IT (Information Technology), ICT (Information and Communication Technology), and informatics usually bear popular understanding of the discipline related to computers and also to the wide range of technologies. The meaning of these terms in education is formally defined based on the context of their use in the curriculum statements.

© Springer International Publishing Switzerland 2015
A. Brodnik and J. Vahrenhold (Eds.): ISSEP 2015, LNCS 9378, pp. 141–154, 2015.
DOI: 10.1007/978-3-319-25396-1_13

The first regular lessons related to 'computers' were held in Poland in the second half of the 1960' when the terms 'computer' and 'informatics' had not been used yet and a computer was a 'mathematical machine', see [19].

In the first informatics syllabus for schools (PTI, 1985) the school subject was called Elements of informatics.

The first national curriculum appeared in the second half of the 1990'. It contained curricula for informatics subjects and the term 'information technology' appeared for the first time as "the combination of informatics technology with other, related technologies, specifically communication technology" (UNESCO/IFIP). Moreover, computers and information technology appeared as educational technology in the curricula of some other subjects. At that time a new term was coined and became popular – 'informatics education' which embraces stand-alone informatics subjects and the use of technology in other subjects.

A word on 'computing' is in place here. In the IEEE/ACM Computing Curriculum 2001 this term embraced 'computer science', 'software engineering', 'information systems', 'information technology'. Since then, computing has changed dramatically and its scope has broadened so much that it is not a single discipline. However it is used in [6] as 'computing education' in a similar sense we use 'informatics education'. This term is also used in the National Curriculum in England (2014) [4] and it covers both areas, computer science and information and communication technology: "The core of computing is computer science ... Computing also ensures that pupils become digitally literate – able to use, and express themselves and develop their ideas through information and communication technology". In Poland, the term 'computing' has no official counterpart – 'informatics' is used for almost everything what is related to computers. However the educational circles popularize the term 'computational thinking', its meaning and importance for the future education of our students, since it has appeared in [23], see also [15].

Recently, when working on a new core curriculum for computer science, to distinguish between any use of computers within informatics education and classes on rigorous computer science we have introduced a term 'computer science education', although school subjects are still called informatics, see Section 3.

Concluding comments on computer terminology in education in Poland today: 'informatics education' refers to any use of computers, informatics, and information and communication technology in other non-informatics subjects as educational tools and methods, and 'computer science education' refers to rigorous learning and teaching of computer science and it also contributes to general informatics education.

1.2 Early History of Computers in Education in Poland

Two initiatives addressed to high school students appeared in two academic centers in Poland, in Wrocław and in Warsaw in the 1960' – see [19] for details.

In 1964/1965, academic teachers from the University of Wrocław initiated informatics classes in two high schools and the subject was called Numerical methods and programming. High school students were writing programs in Algol

60 for numerical calculations during school hours and then ran them on the Elliott 803 computer (made in the UK) installed at the University.

On the other hand in Warsaw, the informatics lessons, which started in 1970, were run in two university mathematics classes in high schools and were devoted mainly to some theoretical foundations of informatics – models of computers and computations. Other initiatives born in Warsaw in the 1970' were: an informatics syllabus for schools and in-service courses for mathematics teachers on teaching informatics in high schools. In the second half of the 1970', informatics was taught in about 1000 high schools in Poland.

It is worth mentioning that the first informatics lessons in schools in Poland were initiated by leading mathematicians and computer scientists of their time working in academic institutions. Teachers and instructors in schools and students involved in the first informatics classes admit today a great concern and engagement of the initiators of those first classes – professors Stefan Paszkowski in Wrocław and Hanna Szmuszkowicz and Zdzisław Pawlak in Warsaw.

The first informatics syllabus was proposed by the Polish Informatics Society (PTI) and then approved by the Ministry of Education in 1985. The school subject was called Elements of informatics. The curriculum covered the topics related to the use of microcomputer applications (for text editing, creating graphics and sounds, building tables and simple databases, making simulations) and elements of structural programming using Logo mainly for drawing pictures and operations on lists of characters.

For the next 10 years (micro)computers were mainly used in teaching informatics as a stand-alone subject and only occasionally they were used for supporting teaching and learning other subjects. Then the development of user-friendly human-computer interfaces and the Internet begun to influence the way computers were used in schools. In the mid 1990', the term 'information technology', as 'informatics for all students', was accepted by the education policy makers in Poland. The emphasis in education has moved from computer science to information technology, from constructing computer solutions to using ready-made tools, from computer science for some students to information technology for all. In [15] we demonstrated however that learning information technology can enhance algorithmic and computational thinking skills in solving with computers problems which arise in various school subjects and other areas.

In the beginning of this century, the national core curriculum contained the following compulsory subjects: informatics (in 4-6 grades in primary school – 2 hours per week for one year or 1 hour per week for two years), informatics (in middle school – 2 hours per week for one year or 1 hour per week for two years), information technology (in high school – 2 hours per week for one year), and an elective subject informatics (in high schools – 3 hours per week for two years). students could also take an external final examination (matura in Polish) in informatics.

In our approach to informatics education we make a general assumption that informatics (= computer science) deals mainly with creating 'new products' related to computers (such as hardware, computer tools, programs and software, algorithms, concepts, theories, etc.) and information technology is mainly using 'informatics (computer related) products'. Although this distinction does not define either informatics or information technology, it is very useful in describing the scope and methodology of learning and teaching both subjects. It is quite important for students' achievement that information technology, especially its sophisticated tools, may be also used to create highly innovative and involved computer products. However, their novelty and ingenuity contribute to the discipline to which they belong, rather than to computer science.

Regarding information technology [9], [10] we convince our students to elaborate her/his style of working with information. Application software programs, such as editors (text and graphics), spreadsheets, presentation programs, usually have several options which support a user in improving her/his style (e.g., styles, templates, wizards, etc.). Elements of style are also very important when working with information on the Internet, and in searching, publishing and communicating on the web.

In teaching informatics (computer science), an algorithmic problem solving approach is suggested for the systematic development of a computer solution for a problem, which covers the entire process of designing and implementing the solution. This methodology is aimed at generating good solutions, characterised by three fundamental properties: readability (the solution is understandable to anyone who is familiar with the problem domain and computer tools used), correctness (the solution satisfies the problem specification), and efficiency (the solution doesn't waste computing resources, such as time and space). The methodology consists of six stages: (1) Analysis of a problem situation. (2) Development of a specification of the problem. (3) Designing a computer solution of the problem. (4) Coding. (5) Testing a computer solution. (6) Presentation and discussion, see [15] for details. These 6 stages of the algorithmic approach are functionally very similar to the stages in the operational definition of computational thinking, developed recently, see [7], [17].

2   Informatics Education in Poland Today

2.1   Informatics Education

For a long time formal education in Poland started at the age of 7, which has recently been lowered to 6. Since 1999 the school system at the primary and secondary levels has consisted of three stages:

- primary school – 0-6 grades (age 6 to 13);
- middle school (in Polish: gimnazjum) – 7-9 grades (age 13 to 16);
- high school – 10-12 grades (to 13 in vocational schools) – (age 16 to 19).

We describe here in more details the present curriculum of stand-alone informatics subjects approved at the end of 2008 and introduced to primary schools (1-3) and to middle schools in 2008 and to primary schools (4-6) and to high schools in 2012. The new curriculum described in Section 3 is in some parts extension and modification of the present curriculum statements towards replacing activities within information technology by learning rigorous computer science, including programming.

Primary Schools

In primary schools a stand-alone informatics subject is now called **computer activities** and runs through grades 1 to 6. In grades 1-3, computer activities are supposed to be fully integrated with other activities like reading, writing, calculating, drawing, playing etc. At the next stages of education students are expected to use computers as tools supporting learning of various subjects and disciplines, formal, non-formal, and incidental in school and at home

Computer activities in grades 4-6 lay down solid knowledge and skills within information and communication technology to be used at the next levels.

Middle Schools

Informatics in middle schools is taught for at least 2 hours per week for one year or one hour per week for two years. The curriculum contains a section on algorithmics, algorithmic thinking and solving problems with computers. Although programming is not included in the curriculum, an introduction to Logo or to another programming language is a popular practice in some schools. Within algorithmics, students are expected to be able to: explain what an algorithm is, provide a formal description (specification) of a simple problem situation and propose an algorithm for its solution; use spreadsheets to solve simple algorithmic problems (e.g. the change making problem); describe, how to find an element in an ordered or an unordered sequence of elements; use a simple sorting algorithm (e.g. by counting); run some algorithms on a computer – either writing a program, using a spreadsheets or running an education software.

Informatics in middle schools is supposed to introduce basic elements of informatics, as computer science, important for at least two reasons: as a starting point for informatics education of all students in high schools and as a pre-orientation for those students who might be interested in choosing a high school which offers a specialization in computer science.

The implementation of the curriculum of informatics in middle schools has some undesirable features – most of the teachers admit that they have no time to cover algorithmic topics. However the truth is that the teachers are afraid of these topics since they are not enough confident in their algorithmic knowledge and skills to touch algorithmics with students who quite often have some experience in programming and running their own programs.

High Schools

In the present curriculum for high schools information technology disappeared as a stand-alone subject and informatics has been introduced in its place, for at least 1 hour per week for one year. In this way, informatics for all students in middle schools has been extended to high schools. This new subject is a continuation of informatics from middle school in the area of problem solving and decision making with a computer by applying algorithmic approach and also may serve as a pre-orientation, intended to prepare students for their choice of future study (e.g. informatics as an elective subject), career and jobs in computing related disciplines and areas.

Informatics (understood as a rigorous computer science) remains in high schools as an elective subject and is taught only in some schools 3 hours per week for two years. Students may also take an external final examination (matura in Polish) in informatics.

2.2 Computational Thinking

Since the first informatics lessons in Polish schools in the mid 1960', algorithmic thinking has been the main approach for problem solving and systematic development of computer solutions of problems coming mostly from computer science and its applications.

A much wider view on computing competencies has been proposed by Jeannette Wing in her seminal paper on computational thinking [23]. Earlier, one of the EU directives suggested that traditional skills for everyone known as the 3Rs (i.e. reading, writing and arithmetic) should be extended to 3R+TI by adding skills in applying information technology. Wing has taken this step further by extending algorithmic thinking and fluency with information technology to competencies for all learners which can be used across disciplines as a computing methodology for solving problems and improve understanding of the role of computing in the modern society. Moreover, computational thinking may also "inspire the public's interest in the intellectual adventure of the field of computer science" [23] and as a result may also encourage and motivate students to consider a future career in computer science related disciplines.

Computational thinking includes a range of mental tools that reflect the breadth of computer science, for example, reduction and decomposition of a complex problem in order to solve it efficiently, approximation when an exact solution is beyond the reach of the computer, recursion as a method of inductive thinking and its computer implementation, representation and modelling some aspects of a problem to make it tractable, and heuristic reasoning to develop a solution of intractable problems.

One can observe the influence of computational thinking on other disciplines. On the other hand, computer scientists' interest in other disciplines is driven by their belief that other scientists can benefit from computational thinking. For instance, in mathematics, as formulated by R.W. Hamming in 1959, "the purpose of computing is insight, not numbers" – note, there were only a few computers in 1959. In [21] we suggest how to extend and enrich traditional topics in school mathematics by applying computational thinking to obtain solutions which are supported by the power of computer science as a discipline and computers as computing tools. Moreover, our approach to deal with topics in mathematics with computational thinking and computing tools contributes to constructionist learning, to learning by doing and making meaningful objects in the real world – here computer solutions. Mental tools used for this purpose include: data representations, reductive thinking, approximation of numerical and intractable problems, recursive and logarithmic thinking, heuristics.

We have adopted computational thinking (see [15]) as the main learning and teaching methodology for information technology as a school subject when it was compulsory for all high school students. A similar approach has been also used in some outreach activities [16] aimed at a better preparation of school students for their future decisions to study informatics related disciplines and to encourage them to consider a future career in computing.

The main difference between using information technology and thinking computationally is in going beyond using information technology tools and information towards creating tools and information. It reminds our distinction between informatics (as creation of programs, computers, theories, etc.) and information technology (as applying informatics tools), see Sections 1.2. The creation of tools (e.g. programs) and new information requires thinking processes about how to use abstraction and manipulate data and many other computer science and computing concepts, ideas, and mental tools of computational thinking.

3   A New Computer Science Curriculum

In this Section we report on an initiative[1] to revise the curricula of informatics subjects (decribed briefly in Section 2) in the Polish National Curriculum so that computer science education will cover all school levels in K-12. Fortunately our job was only to redefine the structure and content of the subject curricula since informatics subjects already cover all school levels in the present curriculum.

In the last decade, several national initiatives have been taken to provide relevant computer science education in schools, for instance in UK [4] and USA [6] – on all levels of K-12 education, and in Denmark [3] and New Zealand [2] only in high schools. Our proposal benefits from these and other initiatives as well as from general considerations such as in [22] and in [12].

In Subsection 3.1 we shortly discuss motivations for our initiative, Subsection 3.2 describes the developed computer science curriculum, and finally in Section 3.3 we comment on some aspects of the new curriculum and activities which will support its implementation in schools.

In what follows when 'informatics' is used it stands for 'computer science'.

3.1   Is Computer Science Education in Crisis?

In the 1980'-1990' computer science was confused with computer programming and, as a result, there was a strong opposition among education policy makers and parents to teaching computer science. They argued that only a few high school graduates would choose a career as a programmer. Even today, many

---

[1] The initiative has been developed by the Council for Informatization of Education, the Ministry of National Education. The authors are members of this Council.

people, among them also teachers and academics, do not consider computer science as an independent science and, therefore, as an independent school subject. Most of them confuse computer science and information technology.

Informatics education in school has not cleared up the myths about computer science and it is again confused with computer programming which has recently become easy accessible even for novice programmers. Students can easily access high-level tools for designing and producing complex applications without any knowledge of fundamentals, such as logic, discrete mathematics, programming methodology, or computability.

Since most students are fluent in using computers to play, search the web and communicate they have no interest in pursuing computer science as a career choice. One of the goals of computer science education should be to motivate students to go 'beyond the screen' and investigate how computers work and how software is designed so they can create their own solutions. Computer science lessons should prepare students for further study instead of leaving them satisfied with the knowledge and skills they have already learned.

The society needs a continuous inflow of good students to be educated and trained as professional specialists for computer related jobs in order to sustain the development and achievements that are necessary to meet the expectations of the information society and its citizens.

The White Paper by the CSTA [13] lists a number of challenges and requirements that must be met if we want to succeed in improving computer science education – our new curriculum has been designed to meet these challenges:

- students should acquire a broad overview of the field of computer science;
- instruction should focus on problem solving and algorithmic (computational) thinking;
- computer science should be taught independently of specific application software, programming languages, and environments;
- computer science should be taught using real-world problem situations;
- computer science education should provide a solid background for the professional use of computers in other disciplines.

3.2   The New Computer Science Curriculum

The new computer science curriculum benefits very much from our experience in teaching informatics in schools for more than 30 years (see Sections 1 and 2). In particular, it takes from the present curriculum the hours assigned to informatics subjects and unifies the names of all stand-alone informatics subjects as informatics. Therefore, according to the new curriculum, informatics is a compulsory subject in primary schools (1-6 grades, 1 hour a week for 6 years), middle schools (7-9 grades, 1 hour a week for two years), and high schools (10 grade, 1 hour a week). Moreover, informatics is also an elective subject in high schools (11-12 grades, 3 hours a week for two years) and high school students may graduate in informatics taking the final examination (pl. *matura*) in informatics.

We have been very lucky that the present curriculum already includes informatics subjects on each education level and we have only had to modify and extend their curricula. The same applies to the hours of instruction. Needless to say that otherwise it would be very difficult to impossible to convince the education authorities that the national curriculum needs such changes in the area which is not of the highest priority on the official agenda, unfortunately.

The new computer science curriculum begins with an introduction explaining the importance of computer science for our society in general and for our school students in particular (see Section 3.1 for some general comments). Then follow the curricula for each level of education. Each curricula consists of three parts:

- Second part is the same in all curricula. It includes **Unified aims** which define five knowledge areas in the form of general requirements – see below.
- First part is a description of **Purpose of study**, formulated adequately to the school level.
- The third part consists of detailed **Attainment targets**. The targets grouped according to their aims define the content of each aim adequately to the school level. Thus learning objectives are defined that identify the specific computer science concepts and skills students should learn and achieve in a spiral fashion through the four levels of their education.

The **Unified aims** are as follows:

1. Understanding and analysis of problems – logical and abstract thinking; algorithmic thinking, algorithms and representation of information;
2. Programing and problem solving by using computers and other digital devices – designing and programming algorithms; organizing, searching and sharing information; utilizing computer applications;
3. Using computers, digital devices, and computer networks – principles of functioning of computers, digital devices, and computer networks; performing calculations and executing programs;
4. Developing social competences – communication and cooperation, in particular in virtual environments; project based learning; taking various roles in group projects.
5. Observing law and security principles and regulations – respecting privacy of personal information, intellectual property, data security, netiquette, and social norms; positive and negative impact of technology on culture, social life and security.

Two very important comments regarding the new computer science curriculum are in order. Although any curriculum defines the aims and targets to be included in any school syllabus, the curricula for particular school levels in the new curriculum contain some optional attainment targets which can be freely added to a subject syllabus or assigned only to a group of students. This is a novelty in our national curriculum and leaves some room for **personalized learning** of gifted students as well as students who have particular interests in specific areas of computer science and its applications (such as mathematics, science, arts).

Personalization in the new curriculum is a means to encourage and motivate students to make **personal choices** of a range of computer science topics and areas in middle and high schools what may lead them towards computer science specialization in the next steps of education and in professional career. Personalization is aimed at increasing students' interests in learning then in studying computer science as a discipline, or at least in better understanding how computers and their tools work and can be used in solving problems which may occur in various areas.

Another facet of personalization is a curriculum of compulsory informatics in vocational high schools for computer, electronic, and electric technicians which is fully devoted to learning computer programming.

It should be noted, that as in the UK curriculum [4], our curriculum recognises the value of computer science as the underlying academic discipline and expects students to understand and use the basic concepts and principles of computer science, analyse and solve problems computationally, programming their solution, on one side (see Unified aims No 1 and 2), and on the other side, students are still expected to apply information technology and to be competent, creative, and responsible users of technology in other school subjects, disciplines, and areas of computer applications (see Unified aims No 2 and 5).

3.3 Implementation Comments, Supporting Activities

Role of Programming. From one hand, A. Perlis wrote in 1962 that everyone should learn to program and Mark Prensky declared a few years ago that "The True 21st Century Literacy Is Programming". From the other hand, we should avoid 'the equation': computer science = programming which is accused of killing interest in computer science among school students in 1990'. Not all students will become professional programmers but writing own programs, individually or in a group, they practice creative and computational thinking, and gain skills of the digital era useful for professional and personal life. They should also get some experience in programming other digital instruments, such as toys, robots, vending machines.

Traditionally, a programming language is a language of computer science in a sense that it is a tool for expressing algorithms and communicate them to computers and also to other programmers. However, we should remember that "informatics should be taught independently of specific application software and programming languages and environments", [13].

In [15] and [17] we have extended the meaning of the terms 'program' and 'programming' to see them in a wider context of using computers to solve problems which are not necessarily algorithmic in nature and introducing all students to computational thinking. There are plenty of opportunities to communicate with a computer by means of programs which are created by other programs, not necessarily writing own programs. The following objects are computer programs: spreadsheet, data base, interactive and dynamic presentation, website, and also documents and graphics and they can be used to 'program' a computer. This meaning of programming with no use of a programming language has a psychological advantage over programming in the traditional sense since the majority of students and their parents consider learning a programming language as the first step to a computer science career, but our goal is only to expose all students to computational thinking. In general, computational thinking is not equivalent to thinking process which leads to computer programming.

The above comments put programming in a right position in computer science education in schools – it is not the only way to communicate with computers and to use them for problem solving. However computer programming can enhance students' problem solving skills in a constructivist way by constructing programs as 'real objects' and then use them in learning by doing (solving problems, making experiments with data, verifying hypothesis, proving statements).

Implementing the curriculum we advise to use programming language for visual programming with novice programmers in primary schools and then to switch for textual language in middle and high schools. No particular language is recommended – the choice is left to teachers and students.

Computer Science Unplugged. Computer science unplugged (CS Unplugged) activities introduce children[2] to fundamental computing concepts and do not require a computer, see [5]. Such activities promote creativity, problem solving skills, and cooperation in groups.

We have extended this approach by introducing some computer activities and integrating them with what children are doing and used it at a children's university to introduce very young students to some concepts in computer science, see [20]. Children work in a number of environments which consist of two stages: first they are engaged in cooperative games and puzzles that use concrete objects (like in CS unplugged), and then they move to computational thinking about the objects and about the concepts they are learning. In this way we introduce our young students to such computer science concepts as: calculations using mechanical tools, complexity issues (the Tower of Hanoi, Fibonacci numbers, and binary search), and graph models (of real world situations). Our approach contributes to constructionist learning, to learning by doing and making meaningful objects in the real world, computational models of real-world situations. Our learning environments are extensions of 'unplugged' ones by encouraging children to purposely and properly use computers for certain activities.

Integration of Computer Science with Other Subjects. We have reviewed the curricula for all other school subject on all school levels and all topics (attainment targets) which are appropriate for augmenting by including and using computer science concepts, skills, and mental tools, have been annotated with

---

[2] The first author of this paper used the CS Unplugged approach in 1970' when introducing the concept of a stable marriage or pairing (due to Lloyd Shapley, the Nobel Prize winner in Economics in 2012) to university students by 'playing' the algorithm with groups of students.

comments how to apply computational thinking to enhance knowledge and skills in the other subjects. We expect a cooperation with teachers of other subjects in implementing our ideas how to integrate computer science with other areas.

Teacher Preparation: Standards, Training, and Evaluation. Preparation of teachers is a crucial factor for the success of introducing the new computer science curriculum to classrooms. The computer science education standards for teacher preparation, which are similar to the ISTE standards [11], have been developed. This standards are operational and include what teachers should be able to do to inspire, motivate and engage students and to promote students' ability to learn effectively. Moreover, the new standards focus also on teachers' engagement in professional development – candidate teachers may come from various pedagogical and subject areas and they may need personalized professional development and training, see [8].

We have also developed a certification procedure, which evaluates in the classroom teacher's preparation for effective and successful managing of learning computer science by her/his students. This procedure is similar to that proposed in [18] for preparation of any teacher to use information technology in the classroom. It is worth to mention that the main purpose of evaluation of teacher's work is to help and support her/him in better preparation for teaching computer science. We are glad to find out that our approach to teachers' certification has many elements in common with the teacher evaluation and improvement system discussed and proposed in the Measures of Effective Teaching Project supported by the Bill and Melinda Gates Foundation.

Outreach Activities. Various outreach initiatives and activities in the area of computer science education are organized in Poland nationwide or locally. They range from formal and informal lectures, courses, and workshops run by public or private institutions. Such activities contribute to school education by increasing motivation and preparation of school students for their future decisions to study computer science or related fields and become computer specialists. Informatics + was one of such projects (see [16]). More than 15 000 students from five regions of Poland took part in this project during 3 years. The project Informatyka + was awarded The Best Practices in Education Award by Informatics Europe in 2013 in "recognition of outstanding European educational initiative that improves the quality of informatics teaching and the attractiveness of the discipline."

Extracurricular activities of students, such as described here and below, promote computer science concepts, involving mainly mental tools of computational thinking. Such activities contribute to formal computer science education as well as to broaden informal and incidental learning.

Competitions. Competitions are typical outreach activities, they are usually run by institutions external to schools. These educational events require knowledge and skills exceeding what is taught in schools. They engage and develop skills necessary in the future professional activities such as: constant self-development, self-discipline, hunger for knowledge, ability to work in a team.

Olympiads in Informatics, which in fact are competitions in algorithmics and programming, attract most skillful students. However, the Bebras competition [1] promotes interest in computer science as well as in information and communication technologies to all school students of all grades. Bebras tasks are on concepts coming from information comprehension, logical and algorithmic thinking, games and puzzles, graphical representations of notions and objects, computer and software functions, etc. They aim at developing computational thinking in the contexts coming from various areas and school subjects.

3.4   The Road Map

The new computer science curriculum as described in Section 3.2 has been accepted by the Ministry of National Education and has been made available to teachers and schools in July 2015. Formally it will be included in the National Curriculum, what needs the Parliament approval, in 2016. In the meantime teachers will take part in various in-service courses on how to develop school syllabi based on the new curriculum and to develop educational materials for their instruction and for students. On the other hand, we expect that computer science departments at tertiary institutions will offer continuous in-service training for teachers, partly in the form of blended learning, based on the preparation standards for teachers, see [18].

References

1. Bebras competition, `http://bebras.org/`
2. Bell, T.: Establishing a Nationwide CS Curriculum in New Zealand High Schools. Comm. ACM 57(2), 28-30 (2014)
3. Caspersen, M.E., Nowack, P.: Computational Thinking and Practice - A Generic Approach to Computing in Danish High Schools. In: Carbone, A., Whalley, J. (eds.) CRPIT, vol. 136, pp. 137-143. ACS (2013)
4. `http://computingatschool.org.uk/`
5. Cortina, T.J.: Reaching a Broader Population of Students through "Unplugged" Activities. Comm. ACM 58(3), 25–27 (2015)
6. CSTA: K -12 Computer Science Standards (2011), `http://csta.acm.org/Research/sub/CSTAResearch.html`
7. CSTA: Computational Thinking Task Force, `http://csta.acm.org/Curriculum/sub/CompThinking.html`
8. Gal-Ezer, J., Stephenson, C.: Computer Science Teacher Preparation is Critical. ACM Inroads 1(1), 61–66 (2010)
9. Gurbiel, E., Hardt-Olejniczak, G., Kołczyk, E., Krupicka, H., Sysło, M.M.: Informatics. Textbook for middle school. WSiP, Warszawa (2009) (in Polish)
10. Gurbiel, E., Hardt-Olejniczak, G., Kołczyk, E., Krupicka, H., Sysło, M.M.: Informatyka to podstawa. Textbook for all students in high school. WSiP, Warszawa (2012)
11. ISTE, `http://www.iste.org`
12. Settle, A., Franke, B., Hansen, R., Spaltro, F., Jurisson, C., Rennert-May, C., Wildeman, B.: Infusing Computational Thinking into the Middle- and High-School Curriculum. In: ITiCSE 2012, Haifa, Israel, pp. 22–27 (2012)

13. Stephenson, C., Gal-Ezer, J., Haberman, B., Verno, A.: The New Education Imperative: Improving High School Computer Science Education, Final Report of the CSTA Curriculum Improvement Task Force, CSTA. ACM (February 2005), http://csta.acm.org/Publications/White_Paper07_06.pdf

14. Sysło, M.M., Kwiatkowska, A.B.: Informatics *Versus* Information Technology – How Much Informatics Is Needed to Use Information Technology – A School Perspective. In: Mittermeir, R.T. (ed.) ISSEP 2005. LNCS, vol. 3422, pp. 178–188. Springer, Heidelberg (2005)

15. Sysło, M.M., Kwiatkowska, A.B.: The Challenging Face of Informatics Education in Poland. In: Mittermeir, R.T., Sysło, M.M. (eds.) ISSEP 2008. LNCS, vol. 5090, pp. 1–18. Springer, Heidelberg (2008)

16. Sysło, M.M.: Outreach to Prospective Informatics Students. In: Kalaš, I., Mittermeir, R.T. (eds.) ISSEP 2011. LNCS, vol. 7013, pp. 56–70. Springer, Heidelberg (2011)

17. Sysło, M.M., Kwiatkowska, A.B.: Informatics for All High School Students. In: Diethelm, I., Mittermeir, R.T. (eds.) ISSEP 2013. LNCS, vol. 7780, pp. 43–56. Springer, Heidelberg (2013)

18. Syso, M.M., Kwiatkowska A.B.: E-Teacher Standards and Certificates. In: Reynolds, N., Webb, M. (eds.) Learning while we are connected. WCCE 2013, vol. 2, pp. 145-151. UMK Toruń (2013)

19. Sysło, M.M.: The First 25 Years of Computers in Education in Poland: 1965-1990. In: Tatnall, A., Davey, B. (eds.) History of Computers in Education. IFIP AICT, vol. 424, pp. 266–290. Springer, Heidelberg (2014)

20. Sysło, M.M., Kwiatkowska, A.B.: Playing with Computing at a Children's University. In: WiPSCE 2014, Berlin, Germany, pp. 104–107. ACM (2014)

21. Sysło, M.M., Kwiatkowska, A.B.: Learning Mathematics supported by computational thinking, In: Futschek, G., Kynigos, C. (eds.) Constructionism and Creativity, pp. 258–268. Österreichische Computer Gesellschaft, Vienna (2014)

22. Webb, M.: Considerations for the Design of Computing Curricula. In: Brinda, T., Reynolds, N., Romeike, R. (eds.) KEYCIT 2014, Berlin, pp. 163–173 (2014)

23. Wing, J.M.: Computational thinking. Comm. ACM 49, 33–35 (2006)

Analyzing the Twitter Data Stream Using the Snap! Learning Environment

Andreas Grillenberger and Ralf Romeike

Friedrich–Alexander–Universität Erlangen–Nürnberg (FAU)
Department of computer science, Computing Education Research Group
Martensstraße 3, 91058 Erlangen, Germany
{andreas.grillenberger,ralf.romeike}@fau.de

Abstract. In the last few years, tremendous changes have occurred in the field *data management*, especially in the context of *big data*. Not only approaches for *data analysis* have changed, but also *real–time data analyses* gain in importance and support decision–making in various contexts. One of the most exciting approaches for processing and analyzing large amounts of data in nearly real–time are *data stream systems*.

In this paper, we will demonstrate how such developments in CS can be introduced in CS education by using data stream systems as an example. We will discuss these systems from a CS education point of view and describe an approach for carrying out data stream analysis by using the Twitter stream as data source. Also, we will show how the programming tool **Snap!** can be extended for supporting teaching in this context.

Keywords: Big Data, Data Management, Data Stream Systems, Twitter, Real–Time Data Analyses, CS Education.

1 Introduction

In modern computer science, a major challenge is to process and analyze large amounts of data. Such *data analyses* are central to *big data*—a topic that is frequently being discussed nowadays, not only in CS and in the economy, but also in politics, society and daily life. Especially, the impact of *real–time data analyses* is increasing tremendously. At the same time, data analyses are hard to notice at all in everyday life, but will become even more important with emerging technologies, like the *Internet of Things* or *Cyper-Physical Systems*, as they will provide many additional data and use cases.

While discussions on data are often focused on storing large amounts of them, for example generated by early data retention projects or intelligence agencies, this aspect is a minor challenge today. Instead, the main difficulty is to process and analyze these growing amounts of data. This leads to a new view on data processing: while traditional data analyses are especially focused on relatively static data, typically stored in a database, today data are rather dynamically changing. Also, traditionally it was sufficient to generate results eventually after capturing the data, but today's analyses are often focused on immediate reactions, like in a tsunami warning system based on seismic sensors. However, CS

© Springer International Publishing Switzerland 2015
A. Brodnik and J. Vahrenhold (Eds.): ISSEP 2015, LNCS 9378, pp. 155–164, 2015.
DOI: 10.1007/978-3-319-25396-1_14

education in this context mainly focuses on storing data in a proper way, often using databases and the relational data model as example, while the aspect of analyzing data is typically left out or only considered marginally (cf. [4]).

In the context of such developments, CS education is confronted with the challenge of keeping track with them and incorporating the basic principles into teaching. However, the complexity of such topics makes this a difficult task. An example is big data: despite its relevance for the student's daily life (cf. [6]), this topic is highly complex and hence appears difficult to include in teaching. In this paper we will discuss one of the emerging approaches for handling big data, *data stream systems*, concerning the new view it brings to data management, as well as its main working principles. Thereafter, we will describe an example of how to introduce this topic in teaching and how to arrange teaching in this context in a student–oriented way using active learning, while also fostering competencies that can be used for overcoming today's flood of information. Additionally, by using Snap! [7] as an example, we will show how an universal programming tool can be extended in order to support teaching of the principles of such new developments.

2   Data Stream Systems

2.1   Function of Data Stream Systems

In modern information systems, the challenge of handling large amounts of data in nearly real–time is becoming increasingly important. While data are typically stored and processed using *Database Management Systems* (DBMS), this approach is being challenged by increasingly large amounts of varying data in short time–spans (*big data*), because traditional approaches can hardly fulfill modern requirements like real–time analyses. Instead, they are designed for use cases in which immediate reactions are not required, such as analyzing business data. But when direct reactions are essential, immediate processing of data is inevitable, e. g. when measuring high values of tectonic movements in a tsunami warning system. While in some cases this challenge might be overcome by accelerating data processing using more powerful machines, this cannot solve the problem in general. A fundamental question in this context is: "*Why do we store all the data?*" In modern *data management*, one approach to address this problem is not to store all the data but only the uncommon ones which have a higher self–information, while presuming that if nothing was stored, everything was as usual. Coming back to the tsunami warning system, it does not make sense to save values of nearly no tectonic movement. While the traditional approach tries to gather as much data as possible on an object/model in order to enable any desired further analysis of them (cf. fig. 1), the approach of *data stream systems* (DSS) is only suitable when there is a clear analysis goal and when criteria can be defined before starting the analysis: they analyze a data stream by filtering out the relevant data on–the–fly (cf. fig. 2). In an analogy, we can describe the database approach as a hamster who collects food on stock, while DSS are characterized by a bear who catches fishes only when he is hungry.

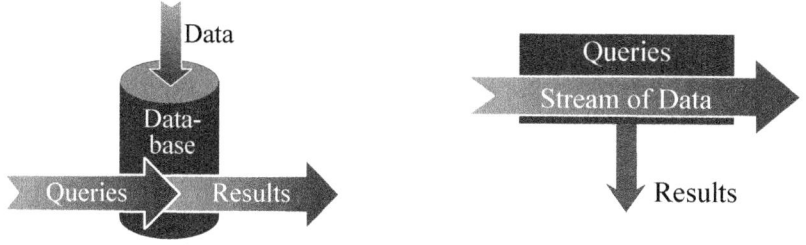

Fig. 1. Function of a database system   **Fig. 2.** Function of a data stream system

The main working principle of data stream systems is to execute queries on *"[. . .] a real–time, continuous, ordered (explicitly by time stamp or implicitly by arrival time) sequence of items"* [3], called *data stream*, instead of one–time queries on stored data. A basic assumption is that when data are arriving in a specified order, each new datum adds new information to the previously received ones or revises them. In consequence, a main characteristic of DSS is that they also produce a continuous stream of results instead of occasional results only when executing a query.

Both, DBMS and DSS, are generally suitable for processing large amounts of data, but they are optimized for different tasks: as databases store all the data for a longer period of time, it is possible to analyze them for correlations or patterns in the data, even such that were unexpected. Such analyses are summarized under the term data mining: *"the process of discovering interesting and useful patterns and relationships in large volumes of data"* [1]. Instead, digging for hidden information without having concrete criteria is not possible using DSS, because the analysis criteria must be defined before the data arrive. For combining the advantages of both systems—immediate reactions, but also long–time availability of all data—using a combination of both types is a promising approach.

2.2   Usage Examples of Data Stream Systems

A typical domain of DSS is monitoring data streams for defined events/criteria. In the following, we will characterize the use of these systems by describing Twitter analysis as example, but most of the described characteristics can also be found when considering other services. While social media play an important role for most students today, they are also a rich data source, e. g. for predicting upcoming trends or for product marketing. We decided for using Twitter as example, as up to 6,000 tweets are posted per second [8] and can be easily accessed using the Twitter API. With each tweet containing not only up to 140 characters as message, but also about 150 additional attributes [2] (e. g. unique ID, author, followers, time stamp, geographical origination, language, information on the profile page of the author), Twitter is a rich information source.[1] When only considering the tweet text (assuming on average 70 characters) and storing it in UTF–8 encoding, this makes about 200 bytes per tweet, in combination with the metadata, a conservative estimation would be around 500 bytes per tweet, which means 3 MB per second or 259.2 GB per day. With these large amounts of data, and especially metadata, various interesting and meaningful analyses can be done: analyzing trends, like Twitter does directly on its start page, reviewing the success of newly released products (even grouped by countries, for example) or generating accurate election forecasts (e. g. [9]). A concrete example is product marketing: when releasing a new product, a typical task is to analyze its success. While such analyses can deliver important results on how to improve the next product, another important aspect is to react to discussions on it. So, an exemplary task is to find regions in which the product needs to be advertised more intensively in order to ensure better success.

Another example for data stream analyses is monitoring of sensors or services like websites. Especially, website monitoring is also highly versatile: e. g. periodically checking if a website is online or offline, monitoring its performance or checking if the price of an item in an online shop has changed. In all these examples, the data source is a continuous stream of data, of which most of the data are not interesting, but the really interesting values can be filtered out efficiently by defining appropriate criteria. A suitable way for depicting the data flow of such analyses is using data flow diagrams. In fig. 3, we depicted such a diagram for monitoring changes on a website.

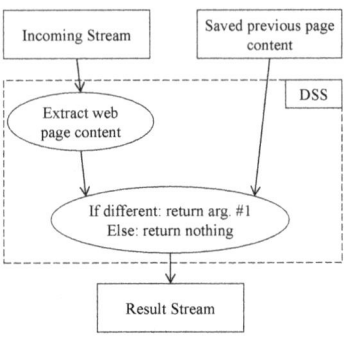

Fig. 3. Data flow of a continuous query for monitoring website changes

2.3 Principles of Data Stream Analyses in Daily Life

The same principle that was used in the previously described examples is also present in everyone's daily life, as they describe the monitoring of data sources, something that everyone does regularly: before refueling our car, we monitor the gasoline price and react as soon as we recognize a cheap price, and when buying goods, we watch the price and buy them when they are on sale. Also, we can use sensors to capture information that is not originally available digitally. This is the case for various innovations in home automation: measuring sunlight for automatically controlling the window shades or measuring temperature for controlling the heater. Even when working in the garden, there are various monitoring aspects, like pouring water on the flowers or the lawn only

---

[1] A complete overview of which metadata a tweet contained in 2010 can be found at http://online.wsj.com/public/resources/documents/TweetMetadata.pdf (last checked: 2015-07-20, created by Raffi Krikorian)

if they are becoming too dry. Systems that are used for the automation of such tasks are often referred to as *Cyber–Physical Systems* (CPS): they perceive their environment using complex sensor structures, react to changes, influence their environment and also communicate with other systems. Such systems do not only enable everyone to monitor and to control their own environment, but they also create new information sources by providing almost every item with its own digital identity—one of the main principles of the *Internet of Things*. This paradigm change in the relationship between physical and virtual objects and practices illustrates the importance of competencies that are needed to cope with the new requirements arising therefrom. Supporting the formation of such competencies is an important task of CS education, which also requires some fundamental knowledge on the basic working principles of such systems.

3   Using Snap! for Twitter Analysis

For CS education, introducing modern approaches like DSS into teaching is a complex task. In particular, topics like big data also cause changes in the relevance of various other concepts, set new emphases and require new examples (cf. [5]). On the other hand, incorporating such topics into teaching enables students to recognize the broadness of CS as well as the chances of using modern CS methods. In this sense, DSS are prototypical for the characteristics of modern data management, e. g. the difference between data and information, the value of data and metadata, standardized data interchange formats, and data analysis methods. With DSS as example, we will illustrate how CS education can support understanding the basic working principles of such modern developments and the acquisition of competencies in this context. For understanding the principles of DSS, students need to be able to use such a system by themselves in order to ensure student–oriented teaching and facilitate active learning. So, in the following we want to demonstrate, how the principles of these systems can be taught using an easy–to–use data stream analysis tool. As typical professional data stream systems are too complex for discovering the main working principles without having in–depth knowledge on the software, we decided to implement a school–focused working example of a DSS based on the easy–to–use programming tool Snap!. In the following we will first describe how this universal programming environment can be used for demonstrating the principles of DSS, in this example using the Twitter stream as data source. Thereafter, we will present the central aspects of our implementation. Our main reason for choosing Twitter is that it offers large amounts of data in an easily accessible way: there are two APIs (one for discrete data, one for streaming data)[2], which are opening up various possibilities. For the described use, mainly the streaming API is of interest: it provides three different endpoints, which means three different data streams. In this example we will access the so–called "filter–stream", which in our tests provided about 16 tweets per second, each enriched with location metadata.

---

[2] The Twitter APIs are described in detail at
   `https://dev.twitter.com/overview/documentation` (last checked: 2015-07-20)

3.1   Possibilities and Usage

Some examples for analyses that can be done using the Twitter data at school
are coming from the previously mentioned context product marketing. For doing such analyses, we need to access the tweet's text, its geographical location,
perhaps hashtags (if provided), retweets or likes and so on. All these information
are provided by the Twitter stream, so we only need to make them accessible
in Snap!. Therefore, we implemented various blocks for accessing and processing
the Twitter data. In the following, we will present a more detailed view on these
blocks, while technical details will follow in section 3.2.

The "get next full tweet" block (fig. 4)
reads the next tweet from the helper via
a HTTP request and returns all attributes
as JSON formatted string. If no value is
returned by the helper, it likely means that
the helper is not running.

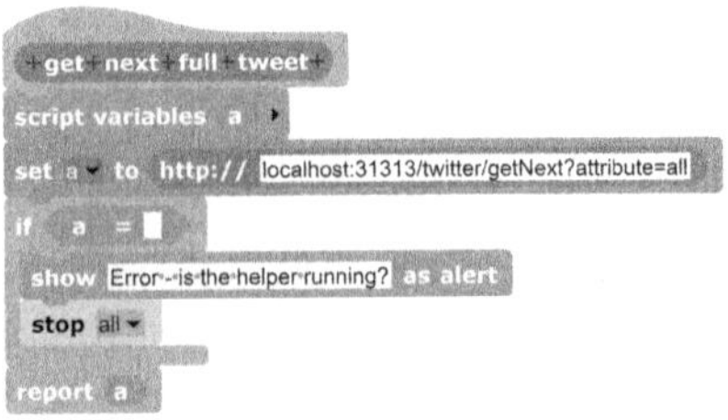

The "read attribute from tweet" block
(fig. 5) typically gets one tweet in JSON
format as argument, which it processes via
a simple JavaScript function, which parses
the JSON string and returns the requested
attribute.

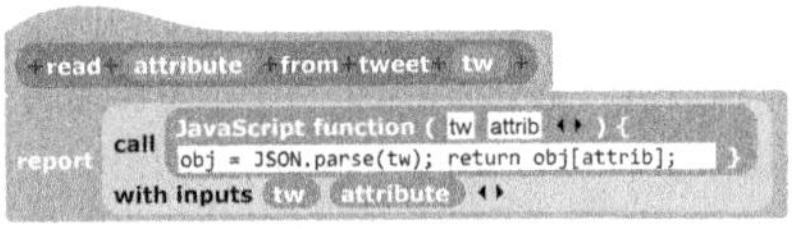

The "for each tweet do" C–shaped block
(fig. 6) is implemented using a forever loop,
in which it reads the next tweet, determines
if its text has at least one character and
then executes the given lambda function
(which is in the user–view represented as all
blocks inside the C–shape), with the JSON
formatted tweet as input.

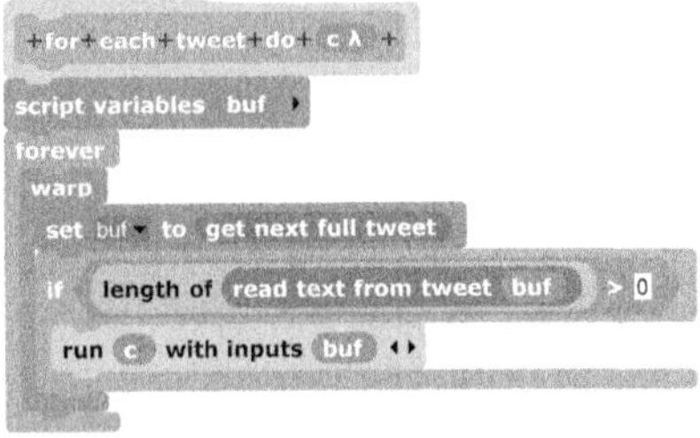

Using these blocks, students can perform
various analyses on tweets, for example
by keyword, language, countries, length of
tweet, hashtags or the color of the author's
profile page. Because only getting numerical and textual results is not a very motivating and interesting outcome, we also
implemented a block for showing the location of tweets on a map. As most simple example, in fig. 7, we have depicted all
tweets that arrived during a time frame of
5 minutes on a map without doing any ad-

Fig. 4. Implementation of the "get
next full tweet" block

Fig. 5. Implementation of the "read
attribute from tweet" block

Fig. 6. Implementation of the "for
each tweet" block

ditional steps. Another visualization that we have implemented, in particular for
doing simple statistical analyses, is a bar chart that can be generated out of a list
of integer values. In fig. 8, we depicted another simple example: determining the
language of the tweet and generating a chart showing the relative frequency of

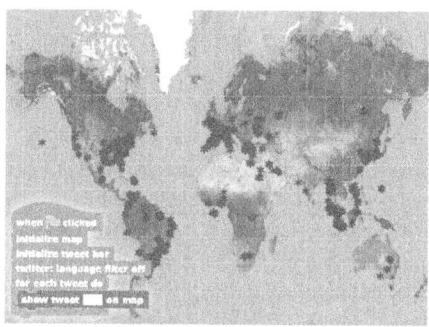

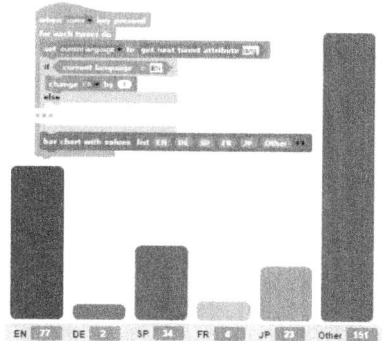

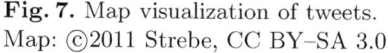

Fig. 7. Map visualization of tweets.
Map: ©2011 Strebe, CC BY–SA 3.0

Fig. 8. Bar chart showing the amount of tweets in different languages.

the languages English, German, Spanish, French, Japanese as well as all others. These visualizations and analysis blocks can be used by students to do analyses depending on their own interests, for example they could analyze which country likes which colors most (determined by the author's profile background color), where a current topic seems to be most discussed or which country favors which stars.

A concrete example for an analysis task in the context of product marketing that can be done at school using our tool is measuring the spread of products in different countries. So, students can e.g. analyze how much the smartphone platforms Android, iOS and Windows as well as their environment are discussed by doing a keyword analysis. By mapping three sets of keywords for these three categories in different colors, the students can get a good overview. However, deeper insights are only possible by using the bar chart and, for example, by restricting the counted tweets to specific countries or regions.

3.2 Realization and Technical Aspects

For implementing the described functions in Snap!, we have chosen an approach that can also be transferred to other data sources (like RSS feeds, website data or sensor data). As both, our approach and Snap! in general, can be used, adopted and extended in a versatile way, in this section we will describe essential details of the implementation that are also relevant for further extensions, using other data analysis approaches and for connecting to other data sources, as well as for transferring our extension to other programming environments. In addition, we will clarify the limitations of our implementation of which the teacher should be aware when using this tool.

As Twitter offers various options for accessing its rich data sources, the first decision was which one to use. We are using the streaming API, which provides three different data streams: the whole stream of data, the so–called fire hose, which is only accessible with special permissions, the sample stream providing a 1% sample of all tweets (about 10 to 15 tweets per second in our tests) as well as the filter stream which lets the user define some criteria (like geolocation or keywords) and returned about 12 to 16 tweets per second. So, not only the slightly higher amount of tweets let us choose the filter API, but especially the fact that we could restrict the results to only tweets containing location data.

After this decision, when implementing our tool we were faced a main challenge: directly connecting from Snap! to Twitter it not possible because of security measures for preventing cross–site scripting attacks in all typical browsers. Hence, we implemented a helper app which acts as proxy between Snap! and Twitter and forwards all tweets to Snap! in a suitable way (JSON format). On the one side, the helper app connects to Twitter using its streaming API, on the other side it offers Snap! the ability to connect to it by running a small web server and allowing Snap! access to it[3]. In Snap!, all functions are implemented using custom blocks and the provided Javascript block, but without modifying the Snap! source code. So, these blocks can be used in the official Snap! installation after importing them.

As a consequence of using a helper application, our implementation cannot fully preserve the data stream character: while there is a typical data stream between Twitter and the helper app, as it is only requested once and then continuously filled by Twitter until the connection ends, we cannot send the data from the helper app to Snap!. So, the "get next full tweet" block breaks up the stream character by requesting every single item on its own. However, as this behavior is hidden from the students, this is only a minor restriction. Also, in order to imitate the stream character as good as possible, we do not cache tweets in the helper app, but instead discard them if there is no incoming request from Snap! in the short time until a new tweet is being received, so if it is not being processed in time. Nevertheless, teachers using this Snap! extension should be aware of this restriction in order to avoid building up misconceptions.

The helper app can also be used with other programming languages and environments that support HTTP requests and parsing text as well as the JSON format, as it provides a universal REST interface and uses JSON for data interchange. So, transferring our solution to e. g. Scratch or AppInventor is possible.

4   Summary

Data stream systems involve various important aspects of CS and in particular of data management. As shown in chapter 2, DSS do not only implement an effective yet easy to understand approach for handling large amounts of data, but they can also serve as an example for data flow modeling, show the principles of real–time data processing and point out the necessity for defined interfaces and exchange formats between systems. Discussing the principles of DSS at school can hence not only show some important principles of CS, but in CS teaching,

---

[3] This technique is known as cross–origin resource sharing (CORS): the target sites allows remote access to its resources by setting a special HTTP header.

it also helps with understanding topics of current interest, like the chances and risks of data analyses.

With the growing importance of data analysis and the large amounts of data that are being generated today, understanding the fundamental concepts and principles in this field becomes increasingly important for handling own data and for understanding common topics in the modern information–driven society. DSS can help students understanding popular data analysis and discussions on data–driven topics, but they also help recognizing the threats accompanying these possibilities. The described tool also gives them the chance to carry out own basic data stream analyses based on the Twitter feed without the need to understand the Twitter API and without possessing in–depth programming skills. Also, this example of analyzing the Twitter data stream can be transferred to many other examples related to the students' daily life: there is only a slight difference to analyzing RSS feeds or other data sources instead of the Twitter stream. As many use cases of DSS are focused on monitoring, this topic addresses another perspective on CS, as students can relate this to their own activities and understand how to automate tasks by using such systems. So, they can also take advantage of this, e. g. by transferring this knowledge to use cases like analyzing the prices of a concrete flight and alerting you when a defined limit is exceeded.

Additionally, incorporating aspects of modern data management into teaching can also foster the formation of various key competencies that are needed for handling own data in an appropriate and responsible way: for example, in the context of DSS, students need to make decisions on whether to store data in a temporary or permanent way, *"understand the purpose of metadata"* and *"combine data in order to gather new information"* [6]. In particular, with the described Twitter analyses, students would also be able to recognize the value of metadata, as most analyses would be relatively meaningless when only considering the tweet text but none of the additional information like the location.

Data stream systems can hence function as an example for involving the ongoing developments and emerging topics of CS into CS education. While current CS curricula do not or only marginally cover such topics, in future the relevance of modern data management topics in CS education is likely to increase: for example, in the context of the web, networking, protocols and so on. In addition, this article demonstrates that considering common tools of CS education from a broader view and using them in a wider context is a promising approach: in this case, the programming environment Snap! could easily be extended to cover aspects of data management and data analysis and to clearly show the main working principles of data stream systems.

References

1. Data Mining. Encyclopdia Britannica, http://www.britannica.com/EBchecked/topic/1056150/data-mining

2. Dwoskin, E.: In a Single Tweet, as Many Pieces of Metadata as There Are Characters. The Washington Journal (2014), http://blogs.wsj.com/digits/2014/06/06/in-a-single-tweet-as-many-pieces-of-metadata-as-there-are-characters

3. Golab, L., Özsu, M.T.: Processing sliding window multi-joins in continuous queries over data streams. In: Proceedings of the 29th International Conference on Very Large Data Bases, VLDB 2003, vol. 29, pp. 500–511. VLDB Endowment (2003)

4. Grillenberger, A., Romeike, R.: A comparison of the field data management and its representation in secondary CS curricula. In: Proceedings of WiPSCE 2014. ACM, Berlin (2014)

5. Grillenberger, A., Romeike, R.: Big data – challenges for computer science education. In: Gülbahar, Y., Karataş, E. (eds.) ISSEP 2014. LNCS, vol. 8730, pp. 29–40. Springer, Heidelberg (2014)

6. Grillenberger, A., Romeike, R.: Teaching data management: key competencies and opportunities. In: Brinda, T., Reynolds, N., Romeike, R. (eds.) KEYCIT 2014 - Key Competencies in Informatics and ICT. Universitätsverlag Potsdam, Commentarii informaticae didacticae (2014)

7. Harvey, B., Mönig, J.: Snap! Reference Manual (2014), http://snap.berkeley.edu/SnapManual.pdf

8. Krikorian, R.: New Tweets per second record, and how (2013), https://blog.twitter.com/node/2845

9. Tumasjan, A., Sprenger, T.O., Sandner, P.G., Welpe, I.M.: Predicting elections with twitter: what 140 characters reveal about political sentiment. In: Proceedings of the Fourth International AAAI Conference on Weblogs and Social Media. The AAAI Press, Menlo Park (2010)

Is Coding the Way to Go?

Violetta Lonati, Dario Malchiodi, Mattia Monga, and Anna Morpurgo*

Università degli Studi di Milano, Milan, Italy
http://aladdin.di.unimi.it

Abstract. Recently, several actions aimed at introducing informatics concepts to young students have been proposed. Among these, the "Hour of Code" initiative addresses a wide audience in several countries worldwide, with the goal of giving everyone the opportunity to learn computer science. This paper compares Hour of Code with an alternative, yet similar, approach which we believe is more effective in exposing pupils to the scientific value of the informatics discipline.

1 Introduction

Programming is at the core of informatics. This is occasionally forgotten in academic circles, but the importance of programming in the intellectual enterprise of computer science becomes evident if one looks at the tag cloud generated by the motivations for the ACM Turing Award winners[1] (see Fig. 1).

Thus, introducing pupils to informatics through programming is a good opportunity to let them bite a real taste of the discipline [9,1]. Certainly this approach represents a better alternative to just exposing the students of primary and secondary schools to the use of computer applications, an unfortunate choice which impacted negatively on computer science education [15]. Driven by this belief, when in 2011 we decided to organize computer science enrichment programs

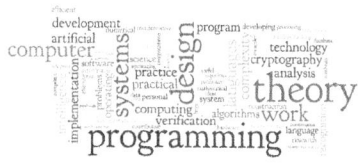

Fig. 1. Tag cloud for ACM Turing award motivations

for secondary schools using the algomotricity methodology [8], we designed some activities precisely focusing on programming, and in particular concerning the problem of guiding an automaton (a "robot": a long standing approach in the teaching of programming [12]) through a simple maze. Since 2013 these activities have consolidated into a short workshop [5] that met the interest of several schools in our town (about 50 classes have participated in two years). We directly conducted the first workshops, and specifically trained some young instructors to conduct them once they were well-established. To assess the outcome of the

---

* The authors would like to thank the 'Istituto Comprensivo Ilaria Alpi', prof. Martina Palazzolo, and all the tutors that helped in conducting the workshops.

[1] Data taken from http://amturing.acm.org/alphabetical.cfm.

A. Brodnik and J. Vahrenhold (Eds.): ISSEP 2015, LNCS 9378, pp. 165–174, 2015.
DOI: 10.1007/978-3-319-25396-1_15

workshops we collected several materials, including interviews with some teachers, questionnaires filled out by pupils, recordings from focus groups. Henceforth this workshop will be referred to as *AlMa* (from "Algomotricity and Mazes"). AlMa is based on mazes, as is the starting example of the "Hour of Code" [11,16], an initiative launched in the US with a strong political endorsement (even President B. Obama wrote a line of Javascript to support it). The mission of the Hour of Code is "every student in every school should have the opportunity to learn computer science"[2]. Hour of Code's introductory proposal (*HoC* for short) is to capture the students with about an hour of coding games. Once students are motivated by HoC, the learning platform can be used (by teachers, but also by students alone) to start looking at more sophisticated computer science concepts and to foster a computational thinking approach. The HoC coding games are exceptionally appealing, featuring amusing characters and fascinating graphics. The platform is based on Blockly [6], a graphical programming environment inspired by Scratch [14]. The HoC website attracted a lot of interest (it claims more than 100 million completions of hours of code) and some positive reports have been recently published [10]. HoC quickly spread to several (180) countries; in Italy HoC is localized as "Programma il futuro" (*"Program the future"*, http://programmailfuturo.it/) and it is driven by the Italian Ministry of Education, Universities and Research. The apparent similarities between AlMa and HoC have forced us to reflect deeply on their differences, and since HoC certainly has a much wider impact than AlMa, we feel the urge to warn the community about what we perceive as a risk to direct pupils, again, towards the wrong target. We believe HoC is a very good intuition and a step that marks an important change of direction in the popularization of computer science; it is perfectly suited to attract pupils, to show them how fun informatics can be, to introduce them to coding. However, if the high-level goal is to show the actual essence and methodology of informatics, and to make students feel its *scientific nature* [7], then HoC can be misleading. Thus, in this paper we will describe AlMa in depth, comparing it with HoC, and we discuss what we believe is working better in our offer. The work is organized as follows: Sect. 2 describes AlMa and Sect. 3 illustrates the assessment of its outcome. Sect. 4 is devoted to comparing AlMa and HoC.

2   The Algomotricity Maze Workshop (AlMa)

AlMa was offered within a wider set of workhops for pupils from 10 to 17 years old. About 50 classes have participated from the end of 2012, from several schools within the Milan District; AlMa was also occasionally proposed in other towns.

Goals. AlMa was developed having in mind the goal of showing the actual essence and methodology of computer science, with the final objective of capturing students to the challenges of a fascinating science, driving them to the *the scientific nature of informatics* [7]. The discipline is introduced in terms of activities focusing on the key themes of processing, automation and information, and promoting

---

[2] http://code.org

the use of some of the methods involved in computer science. From a more technical point of view, AlMa is meant as an introduction to the core of computer programming. Here, the syntactic issues are not the primary concern; instead, the activities are meant to help develop competences related to problem solving, computational thinking, exploratory analysis and scientific research.

Learning methodology. We designed AlMa according to a strategy we call *algomotricity* [8,3,4,5], since the activities focus on *algorithmic* concepts through *motoric* activities, and thus imply a mix of tangible and abstract object manipulations. Algomotricity starts "unplugged" [2] and ends with a computer-based phase to close the loop with pupils' previous acquaintance with applications [13]. This approach gives all participants the opportunity of exploring an informatics concept quite freely and lets them implicitly use the tools of scientific discovery, *i.e.,* formulating hypotheses to be validated by means of experiments. To foster discussion and peer-learning, all activities are performed by groups of pupils.

Description of the activities. AlMa first focuses on the task of verbally guiding a blindfolded person (a "human robot") through a simple path. Working in groups, pupils have to agree on the sequence of statements that a *driver* gives to a human robot. Initially they are allowed to freely interact with the robot, then they are requested to propose a very limited set of primitives to be written each on a sticky note, and to compose them into a program to be executed by the robot. Precisely, they are requested to use at most four different instructions: the constraint is enforced with sticky notes of four different colors at most, and by the requirement that, each time one of a certain color is used, it should always carry the same instruction. Also, they have the possibility of exploiting three basic control structures besides sequence (if, repeat-until, repeat-n-times). Groups may try their solutions as they wish and, when they are ready (normally after 30-45 minutes, depending on the pupils' age and motivation), each group is asked to execute its own program. After pupils have checked that their program allows the robot to correctly carry out the task, the conductor may decide to swap some programs, so that a program is executed by the robot of another group. This allows the instructor to emphasize the ambiguity of some instructions or the dependency of programs on special features of the robot (*e.g.,* step/foot size). In the last phase, which lasts between 20 and 40 minutes, students are given computers and a slightly modified version of Scratch. They are requested to write programs that guide Aladdin[3]'s lamp sprite through mazes of increasing complexity, see Fig. 2. The effect of some commands is illustrated by the conductor: the teacher refers to the first maze in order to show how to use the `move n steps` and the `turn clockwise x degrees` blocks; moreover she or he explains that the "walls" of the mazes are just black regions of the picture: the lamp can walk on them, but a correct solution is one in which this does not happen. Since the full Scratch platform could be confusing for someone who sees it for the first time (and needs to master it in half an hour), our version reduces the available blocks yet maintaining a rich spectrum of possibilities, so

---

[3] Aladdin is the name of our group: `http://aladdin.di.unimi.it`

that the students can focus on the motion and control ones. We also provide just one sensing block: color c_1 is touching color c_2, given with the hint that the lamp has the front in red and the exits have a distinct color, and we briefly explain how the sensor can be used to detect the exit. During the computer part, any working solution is accepted. To promote the use of control structures, a simple competition is proposed: the number of motion blocks used by each team (the lower the better) is recorded on the blackboard.

3 Assessment

To assess the outcome of AlMa, we collected the following materials and analyzed them in the spirit of grounded theory: (1) field notes written during the observation of some classes taking part in the workshop; (2) questionnaires filled out by pupils; (3) three focus groups with pupils; (4) interviews with some teachers. The assessment process involved 150 pupils and their teachers; all pupils attended the same suburban public school, who promoted the participation of all its 6th-grade classes to AlMa. Each of them filled out the questionnaires; the focus groups involved representative pupils from most of such classes.

Questionnaires. Pupils were asked to answer three open questions. (1) What did you like of the workshop? (2) What didn't you like of the workshop? (3) Is there something you feel you have discovered during the workshop?

We analyzed the answers and identified some recurring themes and strong concepts. Pupils claim to like: the fact that the workshop is both amusing and complicated/clever/challenging/engaging; the fact they have created/built something. They feel they have discovered: the importance of thinking/designing/-figuring in one mind's what to do before doing it; the need for precision; that computers and other automatic devices do not work alone, but follow commands; that computer science is not only using computers; that informatics is a science; that informatics may be fun. It is worth noticing that such concepts emerged from all classes quite uniformly, thus they can be considered well-representative of the content and methodology of the AlMa proposal, and not depending on the different conductors or tutors who guided the workshops. We selected the most representative sentences from the questionnaires.

Thinking, designing, mind (answers to question 3) − "To think before doing otherwise you can make mistakes." − "You need to elaborate and set up your mind properly before acting." − "To plan the work in your mind."

Amusing and... complicated, clever, engaging (answers to question 1) − "Very amusing and complex." − "We played, but at the same time we reasoned."

Precision (answers to question 3) − "You have to be precise." − "Technological devices need very precise commands, in order to work properly."

Create, build (answers to question 1) − "When we had 'created' the maze." − "The part in which we had to 'build' a maze."

Focus groups. We proposed as discussion topics the main themes and concepts arising from the previous analysis. In order to activate the discussion, the selected sentences above were handed out and read aloud with the participants. During the discussion most themes were recognized by all the participants. Everybody agreed on the importance of precision to avoid errors and/or risks for the robot, both during the execution of instructions, and when defining the instructions themselves (e.g., how many steps, which turning angle). We registered a unanimous agreement also on the need for reasoning before doing; in the discussions pupils repeatedly used verbs like *thinking, processing, preparing, foreseeing, understanding, solving, schematizing, agreeing*; or terms like *problem* and *logic*; or expressions like *organizing, ordering, putting together*, referred to both *ideas* and *instructions* (in the form of sticky notes or Scratch blocks). They confirmed that the tasks they had to carry out were fun and difficult at the same time, and stressed the fact that the challenge was part of the amusement, because "solving complex tasks is rewarding". However, when asked whether tackling with complex tasks is always amusing, they all clearly gave a negative answer ("I'm willing to use my brain, if the situation is enjoyable."), and pointed out that in this case the activities were fun per se. Words like *playing* or *game* were used to describe the activities, but someone felt such terms too reductive: "it was not child's play", "it was educational". Not everyone acknowledged that during the workshop a creative/building process took place. However, some pupils could establish links: to build is seen as a synonym for *to combine*, or *to put together*, hence this verb is associated with the process of combining sticky notes or blocks in Scratch; *inventing the actions* to write on the sticky notes was experienced as a creative process; for someone, even though the path was prearranged, but groups had to *create* the solution to go through it. Another topic proposed during the discussion is the perceived relationship between the workshop and the subjects taught in school. The concept of *precision* was immediately associated with technical drawing and mathematics; geometry was associated with the measure of length (number of steps of the robots) and turn angles. No spontaneous reference to science emerged. After the moderator suggested some hints, however, all pupils easily associated what happened during the workshop with the typical *observation-hypothesis-prediction-testing-analysis cycle* of the *scientific method*, and in particular with the concept of *experiment*. They told about several episodes when they made a hypothesis (for instance about how many steps were needed), designed a program/experiment, executed/tested it, and verified the correctness of their hypothesis. They also recalled that, when the experiment failed, they reviewed the hypothesis according to its outcome, and started the process anew: "the robot went too far, let's try with fewer steps!" And "when something goes wrong, you often discover something new that you didn't imagine before" (e.g., one is concerned about the number of steps, but finds out that also the turn angle is wrong). Such an approach was also used to choose among ideas proposed by different members of a group: some were tried and failed, while other survived to the experiment and were accepted in the final solution.

4   AlMa vs HoC

AlMa and HoC are apparently very similar. Indeed: (1) AlMa and HoC share the same high level goal: expose to informatics a variety of pupils (not necessarily involved in a computing curriculum), attract them through playful activities, and let them discover how fun and rewarding working with information sciences might be; (2) both are designed as a *first*, short, experience ("an hour" or so: AlMa lasts normally an hour and a half), possibly unrelated to a more structured study of the discipline; (3) AlMa and Hoc are conceived around the same theme: the problem of guiding an automaton through a simple maze. While similar (in fact Fig. 2(h) and Fig. 2(c) represent virtually the same maze), the context in which the tasks are proposed is rather different.

4.1   Algorithm, Program, and Code

To better illustrate the difference between AlMa and HoC, we discuss three terms that are sometimes informally used as synonyms. However their differences, while subtle, are crucial when one has to decide which one has the most potential to attract the creative energies of pupils to our discipline.

Algorithm is possibly the most noble term, with a long tradition (and several formal definitions that here we explicitly ignore): an algorithm is an *effective* procedure to reach, in finite time, a goal[4]. The key point is its effectiveness, a notion that could be clarified only by modern mathematics (Church and Turing above all): Euclid, Fibonacci, and al-Khwārizmī described their famous algorithms *on the assumption that their atomic steps were feasible and sensible.* A **program** is usually defined as an algorithm written in a programming language. In other words, in the post-Church/Turing/Von Neumann era a program is a procedure described in terms of the primitives provided by *a specific interpreter.* As the latter introduces specific syntax and semantics, converting an algorithm into a program can be a complex and creative task, a task largely independent from that of getting to an algorithm solving a specific problem. The recent parlance introduced a third term: **code**. What is then the difference with respect to a program? The word itself suggests a further reduction in the degrees of freedom, a constrained bijection between the procedure one has in mind and its machine implementation. In fact, this word seems well suited when one wants to emphasize the technological context of a program. Coding and programming are sometimes used as synonyms; we surely acknowledge that programming includes a coding activity, but we believe it entails, in general, a more complex endeavor.

4.2   The Hour of Code (HoC)

HoC was launched in 2013 as an activity planned in the Computer Science Education Week, in collaboration with big names of the software industry (Microsoft,

---

[4] One of the most rewarding activities we propose to teachers in our seminars on the didactic of informatics is the discussion of the notion of algorithm: we propose several procedures (cooking recipes, driving directions,...) and we ask why they are or are not actual algorithms.

Google, Apple, Bill Gates, Mark Zuckerberg,...)[11]. Its claimed goal is to "introduce computer programming to all students, to remove the veil of mystery that surrounds the field" [16], by also increasing the participation of women and other underrepresented students to computer science. Although the main HoC offer is based on an online activity, it does exist also in an "unplugged" version. Surprisingly, the unplugged alternative is rather different from the interactive one[5] is rather different: it proposes different tasks focusing more on methodological issues than on coding. To go beyond the first introductory hour, the `code.org` web site proposes also a 20-hour curriculum, with a mix of online and unplugged activities. In fact, it serves a lot of captivating videos and teaching resources, mostly about programming, based on Blockly and Javascript. In this paper, however, we focus only on the introductory part, intended to capture the interest of a vast audience of students who were never exposed to the fascination of the discipline. Moreover, the one hour format makes it comparable with AlMa which has similar goals of letting students meet computer science for the first time. The online HoC is entirely driven by the interactive puzzles: twenty mazes are proposed with increasing difficulties and by changing the constraints and the degrees of freedom of the "robot". The progression is the following (a sample of mazes is illustrated in Fig. 2(f)–(i)): **Mazes 1–5:** the students must code the solution by using the three blocks `move forward`, `turn left`, `turn right`; **Mazes 6–9:** a block `repeat` n `times` is added; **Maze 9:** the solution has to follow a constraint, given as a *grey* block which cannot be moved away; **Maze 10:** a block `repeat until at exit` is added (and the `repeat` n `times` is removed); **Mazes 11–13:** the solution has to use the `repeat until at exit` block, since it must contain no more than four blocks; **Maze 14:** a block `if path to the right` is added: it must be used correctly in a predefined scaffolding of four grey blocks; **Maze 15:** the block `if path to the right` has to be composed to meet the requirement of a solution with less than five blocks; **Mazes 16–17:** the solution should use as few blocks as possible (available: `move forward`, `turn left`, `turn right`, `repeat until at exit`, `if path to the right`); **Mazes 18–19:** the block `if path to the right` has now also an 'else' branch; **Maze 20:** in a predefined scaffolding of three grey blocks two selections are nested: the puzzle can be solved by choosing the appropriate statement to be put in the resulting three branches. Besides the constraints described above, it is also worth noting that each puzzle gives the solver just the subset of blocks useful to each quiz, although the subset is not necessarily minimal: for example, the `turn` block has a parameter `right` or `left`, but the player always has two blocks, one with the parameter set to right and one with the parameter set to left.

4.3 Differences

Problem solving or coding a predefined solution? Solving a problem is always a complex task: one has to distinguish the relevant pieces of information among irrelevant parts and build a model apt to reason about the solution. If a problem

---

[5] http://studio.code.org/s/20-hour/stage/3/puzzle/1

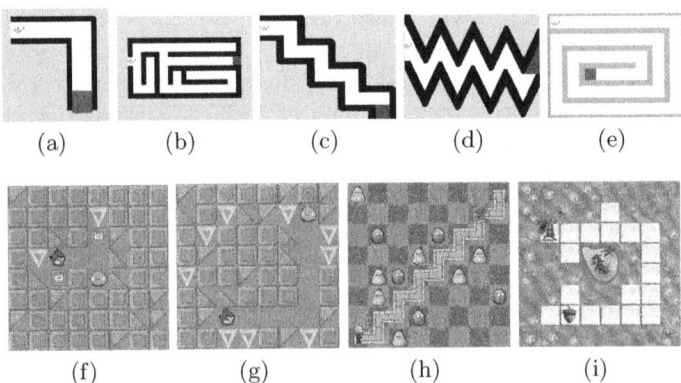

Fig. 2. Mazes used in AlMA ((a)–(e)) and in HoC, taken from http://studio. code.org: maze 3 (f), maze 8 (g), maze 12 (h), and maze 19 (i).

is given without noise, or already abstracted in a specific model, it is probably better considered as an *exercise* in a specific solving technique. In HoC, solutions are predefined (an easy trick to enable simple automatic checking) and almost suggested by the platform itself: the mazes are given within a grid, the robot moves with grid oriented steps, the number of blocks needed is given as a hint. Thus, the solver is left with the exercise of *coding* a solution by choosing the right blocks. When *the* solution is found, the system rewards the solver with a message giving the *number of lines of code* written so far (*"Congratulations! You just wrote 5 lines of code! All-time total: 6 lines of code. Even top universities teach block-based coding (e.g., Berkeley, Harvard). But under the hood, the blocks you have assembled can also be shown in JavaScript, the world's most widely used coding language"* [JavaScript version of the code follows].) Besides noting that most software engineers would agree that the number of "lines" is a misleading metrics, incidentally, it is worth noting that the generated Javascript code reveals some unfortunate design choices: for example, the `turn` block can be used to turn left or right by changing its variable part; instead, the generated code uses two different functions (`turnLeft()` and `turnRight()`), partially breaking the block metaphor and suggesting a questionable programming practice.

   AlMa, instead, proposes little real problems. Students need to "formalize" them, they must find their own way to a solution, not just code a predefined one. Thus, the activity works on the interplay between algorithms and programs. When the pupils drive a blindfolded mate with a finite number of instructions, they reason on what is effective and feasible: and the power of the interpreter is in a large part a choice they explicitly do. Moreover, after the pupils themselves have checked that their program solves the task, when they compare it with others' programs, they discover that some of the assumptions they have made are not valid in the slightly different context of the other teams' settings. Comparing HoC to AlMa, a teacher said: "It lacks the first part which provides the link to reality, to the difficulties of a real problem with its complexities and all its possibilities. Problem solving is the skill to reason computationally."

A riddle or a creative process? In most cases, the solution of HoC puzzles is unique, and the students must aim at guessing it. On the contrary, in AlMa any working solution is accepted. Pupils find out very soon how many different ways there are to accomplish the same goal, a first step in understanding that a program has also *non-functional* properties one might care of. At first, it may seem that students are left alone with Scratch in an intimidating free space of possibilities with just few clues to find their solutions, but the computer game comes after an even more open motoric part in which, however, they *invented* their instructions. What we found is that, after such a step, Scratch's blocks are a quite natural thing to use, and pupils can start trying to find the most similar ones to the commands they conceived in their sticky notes solution.

To be driven or to explore? AlMa aims at giving the students a meaningful problem to be explored in a suitably open context. We provide just a few restrictions designed to support their own inquiry. HoC proposes puzzles whose text mentions enough constraints to rule out all the solutions that do not use the intended blocks, especially when new blocks are introduced. In an interview, a teacher who had tried HoC and AlMa said: "HoC is very constrained, guided, it is more a nice tutorial to the visual framework and the use of blocks, than an actual opportunity for a problem solving activity". While the unplugged version of HoC advertises *computational thinking* as its main goal, the interactive HoC offers very few chances for exploring it. A teacher who proposed AlMa to her pupils last year (6th grade), this year invited them to participate to HoC and she reported they were facilitated a lot by they previous experience with AlMa: "the algomotricity experience introduced pupils to computational thinking, if they hadn't done it I should have figured out some introductory activity before presenting them the HoC proposal". The open-ended activities proposed in AlMa, indeed, encourage the participants to formulate original ideas. Moreover, the team setting forces the pupils to convince the other mates that their proposals work correctly: they need to describe them properly and devise a way to show the correctness of their hypotheses. Thus pupils happen to put into practice and experience the *scientific method*, even though they usually are unaware of this fact. Such an approach is not particularly useful when solving HoC mazes, where an easier (and faster) trial-and-error strategy is generally effective enough.

4.4   Discussion

All in all, if the final objective is to capture students to the challenges of a fascinating science, we believe HoC risks to give an incorrect first impression, only slightly different from the instrumental view so common in the teaching of informatics based on tools and computer applications. With the HoC approach, the scientific nature of informatics is not fully conveyed. Maybe it will be recognized later, if the pupils decide to go beyond the excitation of moving amusing characters. But if we want to show the actual essence and methodology of computer science, why not let pupils enjoy discovering informatics from the beginning?

References

1. Armoni, M., Meerbaum-Salant, O., Ben-Ari, M.: From Scratch to "real" programming. ACM Transactions on Computing Education (TOCE) 14(4), 25 (2015)
2. Bell, T., Rosamond, F., Casey, N.: Computer science unplugged and related projects in math and computer science popularization. In: Bodlaender, H.L., Downey, R., Fomin, F.V., Marx, D. (eds.) The Multivariate Algorithmic Revolution and Beyond. LNCS, vol. 7370, pp. 398–456. Springer, Heidelberg (2012)
3. Bellettini, C., Lonati, V., Malchiodi, D., Monga, M., Morpurgo, A., Torelli, M.: Exploring the processing of formatted texts by a kynesthetic approach. In: Proc. of the 7th WiPSCE 2012, pp. 143–144. ACM, New York (2012)
4. Bellettini, C., Lonati, V., Malchiodi, D., Monga, M., Morpurgo, A., Torelli, M.: What you see is what you have in mind: constructing mental models for formatted text processing. In: Proceedings of ISSEP 2013, pp. 139–147. No. 6 in Commentarii informaticae didacticae, Universitätsverlag Potsdam (February 2013), http://opus.kobv.de/ubp/volltexte/2013/6368/pdf/cid06.pdf
5. Bellettini, C., Lonati, V., Malchiodi, D., Monga, M., Morpurgo, A., Torelli, M., Zecca, L.: Extracurricular activities for improving the perception of informatics in secondary schools. In: Gülbahar, Y., Karataş, E. (eds.) ISSEP 2014. LNCS, vol. 8730, pp. 161–172. Springer, Heidelberg (2014)
6. Google: Blockly (2011), https://developers.google.com/blockly
7. Hromkovič, J.: Contributing to general education by teaching informatics. In: Mittermeir, R.T. (ed.) ISSEP 2006. LNCS, vol. 4226, pp. 25–37. Springer, Heidelberg (2006)
8. Lonati, V., Monga, M., Morpurgo, A., Torelli, M.: What's the fun in informatics? working to capture children and teachers into the pleasure of computing. In: Kalaš, I., Mittermeir, R.T. (eds.) ISSEP 2011. LNCS, vol. 7013, pp. 213–224. Springer, Heidelberg (2011)
9. Meerbaum-Salant, O., Armoni, M., Ben-Ari, M.: Learning computer science concepts with Scratch. Computer Science Education 23(3), 239–264 (2013), http://dx.doi.org/10.1080/08993408.2013.832022
10. Nikou, S.A., Economides, A.A.: Measuring student motivation during "the hour of code^{TM}" activities. In: Proc. of the 2014 IEEE 14th Int. Conf. on Advanced Learning Echnologies, ICALT 2014, pp. 744–745. IEEE Computer Society, Washington, DC (2014), http://dx.doi.org/10.1109/ICALT.2014.218
11. Partovi, H., Sahami, M.: The hour of code is coming! SIGCSE Bull. 45(4), 5 (2013), http://doi.acm.org/10.1145/2553042.2553045
12. Pattis, R.E.: Karel the Robot: A Gentle Introduction to the Art of Programming, 1st edn. John Wiley & Sons Inc., New York (1981)
13. Taub, R., Armoni, M., Ben-Ari, M.: CS unplugged and middle-school students' views, attitudes, and intentions regarding CS. TOCE 12(2), 8 (2012), http://doi.acm.org/10.1145/2160547.2160551
14. Team, M.: Scratch (2003), https://scratch.mit.edu
15. The Royal Society: Shut down or restart (January 2012), http://royalsociety.org/education/policy/computing-in-schools/report/
16. Wilson, C.: Hour of code-a record year for computer science. ACM Inroads 6(1), 22–22 (2015), http://doi.acm.org/10.1145/2723168

Visual Literacy in Introductory Informatics Problems

Françoise Tort[1] and Béatrice Drot-Delange[2]

[1] STEF Research Laboratory, Ecole Normale Supérieure de Cachan, France
`francoise.tort@ens-cachan.fr`
[2] Université Clermont Auvergne, Université Blaise Pascal, EA 4281, ACTé, France
`beatrice.drot-delange@univ-bpclermont.fr`

Abstract. The aim of our research work is to understand reasoning activities of students when they solve Bebras tasks, and especially how they use the diagrams in the solving process. We first need to classify them. This paper gives first results of an ongoing work, of characterization of task according to (i) the types of diagrams and interactive artifacts given in statement and (ii) the way they are explicitly involved in solving process by textual statements of problems.

Keywords: visual literacy, diagrams, problem solving, contest, informatics.

1   Introduction

The Bebras International Contest on Informatics and Computer Literacy addresses pupils grade 5 to 13 with the aim to get them and their teachers interested in typical informatics problems. During a week, in several schools of several countries, pupils try to solve short informatics problems, displayed in the form of online interactive questions.

Problems are designed to meet the dual objective: (1) to introduce the participants to concepts and methods that are typical to computer science and (2) to be within the reach of young pupils who have not any teaching in this area. The second aim could be split into: to be comprehensible without any knowledge in computer science and, on the other hand, to be funny and attractive [2]. Thus, Bebras tasks are not designed to assess knowledge or skills learnt at school, nor to measure well-defined competences but to make pupils discover and explore problems that need algorithm skills, use of specific data representation, etc.

Solving a Bebras task requires reading abilities specific to the written content of a problem statement and abilities in the use of interactive artifacts. In the French contests, very few tasks have neither diagram nor structured data presentation and a majority has several. Moreover, many tasks are like small games in which the interaction schema is used to explore data, change the diagram, or execute a program. Manipulation of diagram and artifact is an important part of the solving process. It is about "Transliteracy", as it crosses media, information and computing literacy. The research

© Springer International Publishing Switzerland 2015
A. Brodnik and J. Vahrenhold (Eds.): ISSEP 2015, LNCS 9378, pp. 175–182, 2015.
DOI: 10.1007/978-3-319-25396-1_16

work, we have engaged since one year in the framework of the TRANSLIT project[1], aims to study the way students solve the problems. We have conducted a qualitative analysis of French Bebras task statements to investigate the type of diagrams and interactive artifacts used, and the way they are explicitly involved in solving process in textual statements.

Section 2 presents research works about images in educational resources, and especially in problem solving that have guided this work. Section 3 presents our methodology, based on qualitative analysis, using Nvivo. Section 4 details coding rules appropriate to characterize use of diagrams and interactive artifacts in the Bebras tasks, and the result of this coding on the French tasks. Last section discusses results and draws perspectives for this work.

2   Roles of Diagrams in Problem Solving

Julo [5] identifies various processes involved in the construction of the representation of a problem. The first step is interpreting and selecting the information that characterizes the problem. A second step is the structuring of the initial representation. The last one is the operationalization process that allows the action. Research in science teaching and learning, especially in mathematics [8] shows that visualizations play a role in the building of a representation of the problem (mental images). They play a role also in operationalization, by the manipulations they allow (diagrammatic and spatial representations). What about Bebras tasks?

In [10], the authors have examined pictures in a corpus of 300 Bebras tasks from Slovak contest. They aim at relating informatics concepts at the core of the task to the picture given in the statements, in order to help Bebras tasks' authors to decide of the usefulness of a picture in a task. They characterize pictures according to: their communication function, their content (computer science concepts, user application, problem setup), and their type. The 'function' characterization is based on the taxonomy of illustration summarized by [1] decorative, representational, organizational, transformational, and interpretational. Indeed, this taxonomy gives a good theoretical framework to make recommendations to designers, but it is not appropriate to focus on the activity of the pupils.

In the domain of mathematic learning, Mottet [6] deal with problems where the image is the base of pupils' solving activity. According to him, effect on learning are not directly linked to images but to the activity performed by the pupil on/with it. He defines 'image situations' as problem solving situations where 'the image are not only to be looked at but imply observable productions, verbal, graphical or even practical productions" (p.19). The three main categories of activities on/with images are: reading (and understanding) the image, modifying it, drawing one.

Resolving a Bebras problem involves information obtained through more than one medium, like diagrams or/and artifacts. In [4] cited by [7], the authors distinguish

---

[1] TRANSLIT is a research project funded by the French government. One of its aims is to understand what competences in fields of media, information and computing constitute a "transliteracy".

scientific diagrams into three categories: iconic, schematic, and charts and graphs, depending on the relationship to the object or situation depicted, in terms of degree of iconicity and abstraction. In all cases, the interpretation of diagrams requires to know the reading conventions, not to fail in solving the problem.

Research has shown the benefits of using diagrams in problem solving. In [8] the author notes that, "in problem-solving situations, when texts are available for comparison, diagrams seem to provide more efficient and enhanced analyses that could also significantly reduce cognitive effort" (p. 212). The limits lie in the familiarity with this type of representation and knowledge of conventions (diagram literacy). The context also determines what and how we see diagrams. Research done about understanding of graphs [3] shows that it depends on numerous factors among which visual characteristics of the graphs but also viewers' prior expertise like graphical literacy skills, explanatory and reasoning skills, familiarity with the content [9], etc.

Bebras tasks are based on underlying concepts and methods that pupils are supposed not to have learnt at formal school. Moreover, diagrams and interactive artifacts are not familiar to pupils. We wonder how much relationships between text, artifacts and diagrams help the student to solve problems, despite the difficulties potentially generated by underlying computer science concepts.

3   Method: Qualitative Analysis of Problem Statements

We conducted a qualitative analysis of the composition of Bebras problem statements in order to characterize all information explicitly given to the reader that could help him solving the problem. We analyzed two points: the types of diagrams and artifacts and their role in the solving activity assigned by the textual statement.

In an exploratory approach, we conducted an inductive coding using NVivo 10 [11]. First, over reading problem statements, we annotated images and portions of text according to their contribution to the above questions using a "code"[2]. If such a contribution had already been coded, we reused the same code, if not we created a new one. We looped over problem statements in order to define more and more precisely the codes. The result was a list of documented codes. Given this list, another researcher applied it to the same corpus of tasks (this time in a deductive way). We compared our results and made adjustments in the definition of codes and coding rules in order to match both coding outcomes. In the same time, we grouped codes onto dimensions, and in each dimension, onto sub dimensions. It results in a categorization that can be compared to those described in the literature.

We coded the 90 problems of French Bebras contests from 2012 to 2014.

4   Coding Rules for Bebras Contest Task Diagrams

The qualitative analysis of Bebras Tasks described in the previous section has brought out three characterization dimensions. This section detailed each dimension, for a full overview see Table 1 in next paragraph.

---

[2] The term "node" is used in NVIVO.

4.1   Dimension A: Type of Diagrams

This dimension characterizes the type of diagrams used in the problem statement. It points out degree of iconicity, of complexity and of standardization of diagrams. The free coding gave about twenty codes, grouped into three sub-dimensions, inspired by well-known categorizations of diagrams [4][7] and data representations.

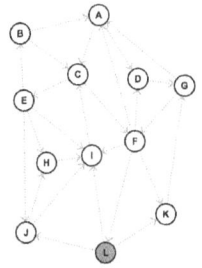

"network graph" - (2014-SP-02)

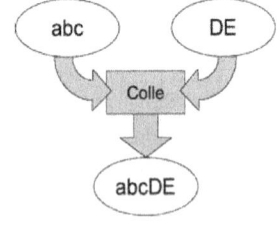

"process diagram" - (2012-SK-02)

"composition of icons" - (2012-AT-10)

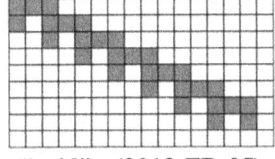

"grid" - (2013-FR-05)

Fig. 1. Examples of diagrams and their coding-in dimension A

Iconic Diagrams: description of concrete objects in which spatial relationships in the diagram are isomorphic to those in the referent object. It covers drawing, screenshot and icons. Icons are simplified drawing that represented real objects by highlighting some of their properties (color, direction, form, parts, etc.). Some objects are compositions of items that are depicted by icons.

Schematic Diagrams: abstract diagrams that simplify complex situations by providing a concise depiction of their abstract structure. In this category, we distinguished: network graphs and trees, grids, maps, diagrams depicting process. Some diagram may be a mix of different types of diagram: some grids contain small icons in cells, some graphs have icons as nodes, some graphs have background drawing, etc. This is not coded explicitly, but gives several codes for the same diagram.

Table and Text Lists: structured presentation of textual information. This covers data tables and lists of keywords, labels or short groups of words.

4.2   Dimension B: References to Diagrams and Artifacts in Texts

This dimension codes the way the textual statement makes an explicit reference to diagrams and artifacts. We got six codes that we classified into two sub-dimensions:

Text Reference to Diagrams: how does the text refer to the diagram? It may give a complete or partial legend: a description of the system symbol used; diagram symbols are designated and linked to the referent objects. It may give instruction explaining how to read the diagram, often with an example; or it may give general description of what the diagram represents.

Legend: "On the diagram, each person is depicted by a circle, persons who know each other are linked by a stroke" (2013-FR-12)
Reading instruction: "Starting at the two wheels at the top, one goes down along the lines, either right or left, to know which elements can be selected and combined." (2012-AT-10)
General description: The drawing below shows a room in which are placed mirrors (2014-FR-05) (The diagram is a grid, some cells contain a slash symbol that depicts the mirrors)

Fig. 2. Examples of portions of text with their coding in dimension B1.

Text Reference to Artifacts: how does the text make a reference to the interactive artifact? It gives instruction on how to interact on the artifact, where to click and how or it gives some general advice on how to use the artifact to solve the problem, like "you can try several times".

4.3   Dimension C: Relationship between Diagrams and the Type of Answer

This dimension codes the relationship between diagrams and the form and type of the expected answer. It is linked to the type of the artifact used to answer. It brings out the nature of the solving activity, by the manipulation of diagrams. The inductive coding has given a dozen free codes, finally reduced to six, grouped into three sub dimensions, corresponding to the types of activities on images proposed in [6].

Diagram Reading: The answer may consist in choosing an element part of a given diagram, via a checkbox list displaying symbols depicting diagram parts. The answer may consist in choosing a diagram among several in a checkbox list. A diagram may be displayed as a model to be reproduced or obtained by an artifact. The given diagram may not be alterable, but the answer needs to read, interpret and understand it.

Diagram Modification: The answer may consist in a program animating a diagram given in the statement; the program may be written in a simple language in an input text area, or it may be given using visual language. The answer may require modifying a diagram by changing the displaying of its elements; by clicking on areas of the diagram, or by using action buttons.

Diagram Creation: In order to cover the three activity types of [6] we added this sub-dimension, corresponding to the design of a new image. But none of the free codes coding the tasks belongs to this category.

The answer consists in filling the pieces with colors.
"modify"   (2013-FR-01)

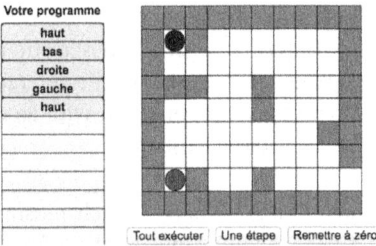

The beads move in the maze when executing the commands listed in the drag and drop list.
"animate" (2014-FR-08)

Fig. 3. Examples of diagrams and their coding in dimension C

4.4   Results: Diagrams in Bebras Tasks

This section describes the result of the coding of the 90 French Bebras tasks (see Table 1). Most include diagrams in problem statement (only 3 over the 90 have none). A majority of diagrams are schematic and give an abstract representation of the problem to be solved. Most of those schematic diagrams are conventional representations used in computer science: graph, tree, grid and also diagrams showing processes (changing states of objects) or structural relationship between objects. However, standard diagrams from 'modeling languages' in software engineering are not present in French Bebras tasks nor screenshots of software interface. Indeed, a lot of Bebras tasks are about graph theory and algorithm on graphs and very few are about use of application software.

Another frequent type of diagram is icon and drawing. Icons represent objects and highlight on some of their properties by using a visual code for properties values. They are useful for comparison of object properties and checking of their compliance with rules. They offer a representation were the student can easily recognize the referent object, but some characteristics are highlighted sometimes in an arbitrary manner (for example, a small triangle depicts a robot and highlight the direction in which it is oriented). In this sense, those icons are closed to schematic representation.

A schematic diagram uses arbitrary symbol system of representation and well-known schematic diagrams use a standard one. In Bebras tasks, few schematic diagrams respect a standard representation. Yet, a legend or an explanation of how to read the diagram is not always given in the problem statement. The interpretation of the symbol system is part of the problem. Indeed, it is usual, in computer science, to choose representation and decide if they fit the data and help to solve the problem.

Table 1. Result of the coding of the 90 French Bebras Tasks (a task may be coded several times)

A.Types of Diagrams	
Iconic (*47 tasks*)	Drawing, Screenshot, Icons, …
Schematic (*55 tasks*)	Graph or tree (*19 tasks*), Grid or map (*16 tasks*), Process Diagram (*16 tasks*),
Table and text lists	Table and Text lists (*12 tasks*)
B. References to Diagrams and Artifacts in Text	
to Diagrams	Legend (*17 tasks*), Reading Instruction, General description (*29 tasks*)
to Artifacts	How-to use instruction (*38 tasks*), General advice (*13 tasks*)
C. Link between Diagrams and answer	
Reading	Choose an element (*15 tasks*), Select a diagram, Replicate a diagram, Think on
Modification	Animate, Modify (*31 tasks*)
Creation	*(0 task)*

In a majority of French Bebras tasks, interactive artifacts are not simple form fields. It is a complex artifact with many clickable areas in the diagram, drag and drop list of icons or short texts, input text for entering parameters or program text and execute buttons, a result-display area. In one-third of the tasks, the answer consists in modifying a diagram given in the statement, either by clicking on it or by a less direct mean where a click in one area impacts another area. In some cases, the modification gives information useful to solve the problem. Most of the time, the text explains how to use it (where to click or drag and drop), but not how it works and not how to solve the problem with it. It means that understanding the interface of the artifact is not part of the challenge and should not be an obstacle. In contrast, understanding the behavior of the system (the algorithm beside) is part of the problem.

In Bebras task, diagrams and artifacts have a central role in the statement: they may be an input of the solving process, or its final product. Understanding and interpreting diagrams are at the core of the solving process, whatever the answer requires: a direct manipulation, or a mental process. In this sense, Bebras tasks are 'image situation' as defined in [6], where 'situation' means statement for a problem solving. Most of diagrams in Bebras offer heuristic tools that help student make hypothesis about how to search for the solution.

Are the students familiar with the diagrams and the interactive artifact used in Bebras task? We don't think so. Trees are manipulated by pupils at primary school, but only in specific situations, like genealogical trees. Graphs are not part of the curricula. Conventional schematic diagrams are used in technology lessons. The issue is to understand how much manipulations of diagrams help them to understand the problems and to find ways to operationalize the resolution.

5   Perspectives: Next Steps

This paper presents the very first stage of a work in progress. Next steps deal with the observation of students resolving Bebras tasks. We already collected screen video captures and audio of 30 high school students saying aloud what they were doing or having in mind when solving Bebras tasks. The characterization of the role of diagrams and interactive artifacts in problem solving will help us to question and interpret the observations. To investigate the process by which pupils use the artifact in relationship to what they aim to do will certainly be fruitful. Are they an obstacle or a help during the solving process? We will look carefully to the diagrams students draw themselves on paper, and try to characterize them in relationship with our coding system. We also plan to complete the characterization by taking into account other dimensions such as the computer science domains, notions and methods.

References

1. Carney, R.N., Levin, J.R.: Pictorial illustrations still improve students' learning from text. Educ. Psychol. Rev. 14, 5–26 (2002).
2. Dagiene, V., Futschek, G.: Bebras international contest on informatics and computer literacy: Criteria for good tasks. In *Informatics Education-Supporting Computational Thinking* - LNCS 5090, 19–30. Springer (2008).
3. Glazer, N.: Challenges with Graph Interpretation: A Review of the Literature. Studies in Science Education. 47(2), 183–210 (2011).
4. Hegarty, M., Carpenter, P. A., & Just, M. A. Diagrams in the comprehension of scientific texts. In R. Barr, M. L. Kamil, P. Mosenthal, & P. D. Pearson (Eds.), Handbook of reading research. 2, 641-668. NY, NY: Longman (1991)
5. Julo, J.: Représentation des problèmes et réussite en mathématiques. Un apport de la psychologie cognitive à l'enseignement. Presse universitaire de Renne.(1995)
6. Mottet, G.: Les situations-images : Une approche fonctionnelle de l'imagerie dans les apprentissages scientifiques à l'école élémentaire. Aster, 22. (1996).
7. Novick, L.R.: The Importance of Both Diagrammatic Conventions and Domain-Specific Knowledge for Diagram Literacy in Science: The Hierarchy as an Illustrative Case. In D. Barker-Plummer, R. Cox, & N. Swoboda (Éd.), Diagrammatic representation and inference, 1–11. Springer Berlin Heidelberg (2006).
8. Rivera, F.D.: Toward a Visually-Oriented School Mathematics Curriculum: Research, Theory, Practice, and Issues. Springer Science & Business Media (2011).
9. Shah, P., Hoeffner, J.: Review of graph comprehension research: Implications for instruction. Educational Psychology Review. 14(1), 47–69 (2002).
10. Tomcsányiová, M., Kabátová, M.: Categorization of Pictures in Tasks of the Bebras Contest. In: Diethelm, I. and Mittermeir, R.T. (eds.) Informatics in Schools. Sustainable Informatics Education for Pupils of all Ages. LNCS 7780, 184–195. Springer Berlin Heidelberg (2013).
11. Welsh, E.: Dealing with Data: Using NVivo in the Qualitative Data Analysis Process. Forum Qualitative Sozialforschung / Forum: Qualitative Social Research. 3(2). (2002).

Author Index

CPSIA information can be obtained
at www.ICGtesting.com
Printed in the USA
LVOW04s1836171115

462997LV00004B/105/P